Yearbook 2007

LAURENCE KING

Published in 2007 by
Laurence King Publishing Ltd
361–373 City Road
London EC1V 1LR
United Kingdom
email: enquiries@laurenceking.co.uk
www.laurenceking.co.uk

A catalogue record for this book is available
from the British Library.

ISBN-13: 978-1-85669-516-9
ISBN-10: 1-85669-516-6

Guest Editor: Patricia Urquiola
General Editor: Jennifer Hudson
Designer: Philip Lewis
Project Manager: Anne McDowall
Assistant Editors: Anna Frohm and Lindsay Zamponi

Printed in China

Pages 2–3:   Armchair, Log
             Patricia Urquiola
             Artelano, 2006

Page 5:   Side tables, T-table
             Patricia Urquiola
             Kartell, 2005

Pages 6–17: All furniture designed by
             Patricia Urquiola

Page 19:   Cantilever Chair
             Pearson Lloyd
             PA, polycarbonate, PP, steel
             H: 80.6cm (31¾in)
             W: 57cm (22½in)
             D: 45.7cm (18in)
             Allermuir Ltd (part of the Senator
             International Ltd group), UK
             www.allermuir.com

Page 91:   Hangable Scour-pad Glove
             Paolo Ulian
             Microfibre fabric
             W: 20cm (7⅞in)
             L: 27cm (10⅝in)
             Coop, Italy
             www.coop.it

Page 141:   Pendant light series, Beat Light
             Tom Dixon
             Hand-beaten brass with black patinated finish
             Beat Shade-Wide: H: 16cm (6¼in),
             Diam: 36cm (14⅛in)
             Beat Shade-Fat: H: 30cm (11⅞in),
             Diam: 24cm (9½in)
             Beat Shade-Tall: H: 36cm (14⅛in),
             Diam: 19cm (7½in)
             Tom Dixon, UK
             www.tomdixon.net

Page 171:   Candelabrum, Mistic
             Arik Levy
             Boras silicate glass
             H: 20cm (7⅞in) or 45cm (17¾in)
             Gaia & Gino, Turkey
             www.gaiaandgino.com

Page 199:   Double knit, Washingsymbols (W07-7812)
             Naomi Sara van Overbeeke (pattern),
             Jacqueline Janssen, head designer (fabric)
             30% Softcell, soft cotton, 20% polyester,
             15% viscose, 20% polyamide
             W: 230cm (91in)
             Prototype for Moroso
             Innofa, The Netherlands
             www.innofa.com

contents

skin

Lately, I've been fascinated by the concept of skin. Skin is subtle. Skin is clear. It's container and substance at the same time. It's the surface, the impression, the outer world. After years of putting function first, it seems that design is becoming more subjective, and adaptive to our needs, desires and pleasures. I see a focus on textures, patterns, surfaces, coverings in both design and architecture, and in a relatively new genre that is a combination of the two. I am excited by the potentiality of mixing art and craft techniques with modern technologies to achieve a blending of the new and advanced with the traditional – an emotional element that is connected to something we know and recognize but that has been adapted in an innovative way.

However, the danger is that too many possibilities can result in confused or over-personalized designs which become excessive and ultimately uninteresting. It is immensely important to correlate the inside (the function) of a design with the outside (the narrative element or the decoration) – a process analogous to the human body and the delicate and balanced relationship between bone, muscle, skin and nerve. The result can be something sophisticated and beautiful or inharmonious and deformed.

Making design choices – or, indeed, singling out products for this yearbook – is like the process of selecting music on an iPod. Two modes are possible: the linear – searching by song title, keeping to what we know, playing safe, using our memory in a very restricted way; or the arbitrary – selecting the random button and leaving it to chance. The first is predictable, while the latter is exciting and liberating and can lead to something wonderful or to disaster. Of course, even a 'random' selection will be dependent not only on chance but also on ourselves, on how much taste and thought we've put into the pre-selection, and on how lazy or diligent we've been in maintaining the database of music. The way we erase and save certain 'musts' onto our computers will determine the eventual outcome. Choosing only esoteric music will limit the selection, as will picking only a few very commercial songs.

Continually to download haphazardly would be confusing; a certain processing of the selection is necessary. We have to re-examine what's on offer, then reuse and reinterpret it to give it a new identity. In doing so, we run the same risks that we do when listening to a cover version of a favourite song: we might find it disappointing or even painful,

but equally we might discover in it something delightful or even great.

Waiting for new typologies that reflect the evolutionary changes in function or living patterns is boring. But to create something that gives pleasure is extremely gratifying, especially when that pleasure is shared by millions of others. In the creative field there are no universal laws; everything is subjective. In my selection of designs for this yearbook, I am offering you my 'iPod favourites'; I am sharing with you some of the themes, influences and researches that I have enjoyed.

The selection of the yearbook is determined by a specific time frame and by certain pre-qualifications, which for many years have been dominated by Western designers and companies that have access to production networks, fairs, exhibitions and design prizes (with the odd anomaly that finds its way in). I'm hopeful that, in future compilations, a few hits on 'random selection' will add to the richness of designs – and our enjoyment of them – with new input from emerging countries in the Far East, South America and Africa, and with new means of production, distribution, marketing and information. With less pre-selection and fewer filters, there will

Foreword by Patricia Urquiola

Opposite:
Chaise longue, Antibody
Moroso, 2006

Easy chair, Pavo Real
Driade, 2006

be more space for freedom of expression and talent to be shown.

I hope that you will find some 'songs' that you did not know or 'old songs sung to a different tune', and that, whether you like or dislike them, you will find the selection refreshing. It is not an easy balance to maintain. As in many areas of life, it is difficult to gauge what is right for the time; it's easy to be either too late or a little premature. In any case, it is impossible – not to mention boring – to try and please everyone.

Let's enjoy diversity without feeling the need to try to 'catalogue the world' or be inhibited when we discover we share the same direction or emotions as others. I do not believe in particular styles or schools of thought, but I think I have an inherent feel for what is contemporaneous and modern. I recognize waves of influence, or propagation, but feel that it's the way in which these are circulated that makes for richer and more intriguing design pieces.

Choosing products for the yearbook is a great responsibility. Unlike other forms of communication, such as the internet, which isn't printed and can be altered, the yearbook is a direct conduit between what's on offer, the process of selection,

and the final printed form which can have no changes made to it. But this very 'limitation' is part of its beauty. In fact, my hope is that the book will be used in conjunction with both static and dynamic media, that you will be enthused by the images to such an extent that you will want to log on to websites to find out more about a particular product, its designer or manufacturer.

When making my selection, I wanted to reflect the predominant examples of the figurative as opposed to the decorative. I was at first unsure whether to underline the trend or to downplay it, but it's an element in design at present in which I am interested and also one I see as prevalent in not only the work of my generation, but also that of new designers emerging onto the scene: the sons and daughters of the 'abstract' designers that have gone before.

It is a direction that I pursued in an exhibition at Verona's 'Abitare il Tempo 2006', entitled 'Donkey Skin', a sort of auto-ironical inner view of the process of creation. The title of the exhibition comes from Charles Perrault's fairy tale of the same name, in which the donkey's skin becomes a metaphor for the outer appearance we adopt to protect ourselves from unwanted attention,

leaving us free to use our hidden qualities without detection. It deals with the surface of objects, with the interaction between architecture and graphics, with the relationship between two complexities: craftsmanship and technology.

My Milanese training always emphasized the importance of a coherent synthesis; that is, the focussing on the essential of a project, not adding anything that is not needed. With this as a base, I want to debate whether it is possible to combine craft with industry, whether complexity and seriality can co-exist. Industrial design will always be repetitive, but how industrial and how repetitive? It's an open discussion, but whatever conclusion is reached, it is important that nowhere should design become confused with art. Both great in their own way, they are different disciplines that reside on two separate peaks of Mount Olympus; they influence, but should never be mistaken for, one another. The risk in evaluating a possible link between design and craft is that the graphic may become the focus of attention rather than other defining elements. The debate should never lead to a series of beautiful objects that don't function, whose value lessens when they are used rather than just admired. The figurative or decorative

WC and bidet, Pear
Agape, 2004

Salt and pepper pot,
Sergeant Pepper
St. Lorenzo, 2003

element should not be utilized as a short cut to avoid technological issues nor at the price of reduced practicality.

The solution to this problem is to create a link between the great (but in danger of extinction) generation of technicians who work for the major manufacturing companies and, with a love of the products, use their skills to play with surfaces and tactility to get the best possible results, and the new designers, who use computers to create virtual realities. Both are blessed with their own special talents. The technicians see an object in their minds and express it directly with their hands, solving problems through a filter of experience and acquired knowledge, while these digital designers call upon 3D, with its many unexplored possibilities, creating new problems rather than solving them and bringing to the table new challenges and visions with undefined boundaries. Between the two stands the designer of today, whose task it is to harmonize the flow of communication, pushing the limits of the technician while refining the choices of the digital specialist. The result should combine the characteristics of desire, dream and imagination with practical functionality and the pleasure of touch. Semiotics and narrative have

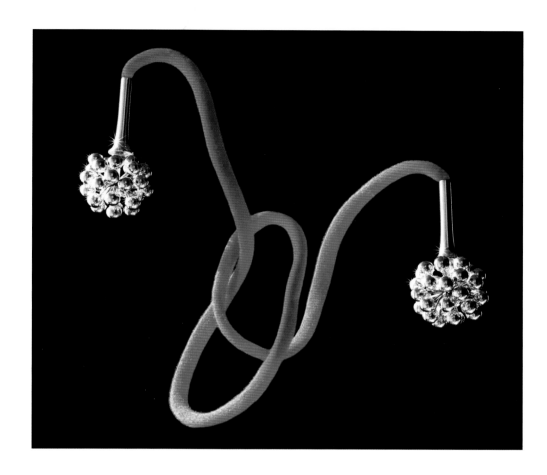

Necklace, Pompom
St. Lorenzo, 2004

Armchair, Smock
Moroso, 2005

9

valid roles to play, but in the world of industrial design have no place without meaning. *Aliquid stat pro aliquo*. A signifier can be transformed into something different only if it has an origin.

I hope that the use of the figurative will result in an output of work that will become neither irritating nor saturative, which has tended to be the case with the overuse of decoration. The key is balance, restraint and the respect of proportion. In the fields of interior design and architecture, this means the consideration of the inner space with respect to the architectural code of the building and also within its urban context. It means re-examining a discipline that desires to transform an existing space into something which relates to the different dimensions surrounding it, making decoration a part of a holistic and organic project. To achieve a project that has unrealizable expectations cannot be considered successful, only unresolved. Offering various tools that can be adopted in different ways and for different uses means that we are less likely to live in a monochromatic, monotonous and repetitive, mass-media-led world. It lessens the possibility of repetition, of the same concept being used universally and out of context, and creates a greater diversity. Design should work in all

directions and should deal with opposites, using the digital in an invisible yet formative way as well as technology from the past innovatively and on a different scale, constantly re-evaluating the weaknesses and limiting the strengths.

Another important element of the design process involves thinking about how a product is used, which means looking not only at materials and techniques, but also at the way in which an object will be seen. A cheap product, if beautiful and in keeping with other objects, can be used in a high-budget project and should not be categorized by its price. A piece designed primarily for a professional context can be used in the home, and a domestic item in the workplace. An expensive object can be used in a public space where it can be admired by many, or we can derive pleasure from looking at it reproduced in the pages of a book or magazine. Property shouldn't equate with pleasure; there is no need to own something to receive happiness from it. Let's enjoy the freedom of mixing the cheap with the expensive, the old with the new, the serious with the ironic, or even something beautiful with something ugly – the juxtaposition could result in a positive outcome. Combinations and risks are infinite and rewarding.

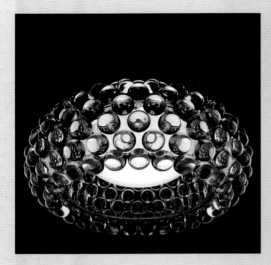

Pendant light, Caboche
Foscarini, 2005

Modular seating system, Lowseat
Moroso, 2001

'I needed space, distance, history and danger, and I was interested in the living world.'

The Coast of Incense, Freya Stark, 1953. Explorer and writer (1893–1993)

Patricia Urquiola is a living dynamo. It's hardly surprising that she has earned the nickname of 'Hurricane'. She takes over any room she walks into with her extraordinary passion and vivacity. With her sexy, guttural, Castilian accent she slips unconsciously from English to Spanish and Italian and the walls vibrate with the sound of her laughter. Her phone is constantly ringing and, with the perfect ease of the contemporary woman, she alternates from giving instructions to colleagues and manufacturers to purring lovingly to her daughter or whispering sweet nothings to *mio amore*. Her sentences are peppered with self-assured exclamations: *ecco*, *capito* and *comunique*. Yet her soft and sensitive side is never far from the surface. She didn't want to make the yearbook selection without Alberto, her partner in life and business associate, and, when he couldn't make it, brought along her friend and fellow designer Christophe de la Fontaine, whose opinions she respects and humour she shares. They constantly sparked off one another, but when decisions had to be made, Patricia, with her strong character, knew exactly what she wanted.

Urquiola is half Basque and half Asturian and spent her teenage summers in the laid-back hippy environment of Ibiza in the '70s. She studied architecture in Madrid, abandoning the course for love and a move to Milan, where she enrolled at the city's polytechnic, graduating in architecture in 1989. Her thesis was supervised by Achille Castiglioni, who became her mentor and who persuaded her to follow in his footsteps, abandoning architecture for design. Under Castiglioni's influence, she learned to focus on the essential, rejecting the superfluous and the fashionable in favour of the classical. He also taught her to be creative in a domestic context, emphasizing the importance of the everyday items that surround us. 'After you meet Castiglioni, you cannot think of anything except how important it is to design. He shows that the little things can be just as interesting as architecture.' She later worked for Vico Magistretti, de Padova and Lissoni Associati, building up a strong head for business and a thorough knowledge of the manufacturing industry.

Her life informs the development of a personal style, which is a perfect blend of simple, clean Italian design subverted by a Spanish flamboyance – visual elegance and technical virtuosity combined with a streak of the anarchic. Her work is at once formal yet cosy, contemporaneous but not fashionable, signature without screaming design, commercially sound while displaying tiny details that have an emotional kick and provoke discussion. And even when playful, fun and feminine, her designs are always functional and on the consumer's side. On the acquisition of one of Urquiola's chairs, Paola Antonelli, MOMA's curator of architecture and design, was quoted as saying 'Patricia is able to create things that are completely innovative, yet perfectly attuned to people's homes.'

It's this blend of being able to 'play the game' while at the same time contradicting it that is one of the reasons Urquiola has made such an impression on the Italian design industry in such a relatively short time. Who at the turn of the century had heard of Patricia Urquiola? Yet since setting up her own studio in 2001 she has become ubiquitous, designing furniture for all the major Italian manufacturers, especially Moroso, with whom she has forged an important relationship. Its creative director, Patrizia Moroso, now a firm friend, recognized in Urquiola what she later referred to as a 'perfect distillation of everything Italian'. Moroso offered Patricia the chance to experiment early on in her career, and, within the context of commercial design, to give full reign to her sensuous, passionate and emotional Spanish side. In return, Urquiola has

Introduction by Jennifer Hudson

Opposite:
High-backed chair, Lazy
B&B, 2004

Watch, Buckle
Alessi, 2005

Table with storage
B&B, 2002

Side table, Digitable
B&B, 2005

Side table, T-table
Kartell, 2005

created many of the company's best-selling products, most notably the Fjord range, the Lowland sofa and the Bloomy family of chairs.

As well as furniture, Urquiola is presently producing rugs for Paola Lenti and taps for German manufacturer Hansgrohe. She is even turning her hand to her first love, architecture, with villas in Punta del Este, Uruquay, and Udine, Italy, the B&B Italia showroom in Barcelona and a residential tower in Shanghai. Unlike many names that the fickle design press promote one season and cast into obscurity the next, Urquiola's pedigree is a result of hard work and determination and will ensure her a place among those set to stay. Her public image has been built up gradually; it is not dependent on press hype – and as such will not be tied down to a period or nailed to a trend.

Urquiola is among a handful of women who have succeeded in the furniture-design business. Of her own admission she is tenacious and stubborn – characteristics she has needed in order to flourish in an industry where male architects talk to male-run companies with male technicians – yet she maintains a femininity and a flirtatiousness that she is not afraid to use to get what she wants. She is aware that a direct, businesslike approach does not work for a woman in a world dominated by men, and that a more subtle approach is called for. 'Maybe the rules of confrontation regarding egos, language and communication are not yet set for the female designer category, so we need to invent them and bypass the obstacles, without changing ourselves.' Urquiola believes that her femininity has given her an advantage. She is very much a woman of today, with two beautiful daughters and a secure and happy family life, and she has learned to juggle successfully her roles as mother and professional designer. This, she believes, has given her the freedom to embrace a more flexible approach that is so important and useful to a designer.

Patricia is an industrial designer. She works under commission. Communication is all-important to her and is a word that she uses frequently. She collaborates with her clients and their technicians and, guided by each manufacturer's individual specialisms, creates products that are a synthesis between her original concept and what will be demanded by their marketing departments and the needs of the end user. She seeks out the fresh and is influenced by everything from art to everyday life, from travel (which is one of her biggest passions), to memories of childhood, but is convinced that design has to be reproducible and also fit into a context. 'I'm more of an architect than a designer. I put each product in a specific context, *capito*; I never draw just a chair or a lamp. Many times I sense the entire space in which it stands, as well as other objects to which it relates. What is important is the language of the whole.'

Although presently investigating to what extent craft can be combined with industry and seriality, she constantly has in mind the industrial process and the need to remain essential and clean. Her constructions are straightforward, but the unexpected arrives in her application of surface – or what she refers to as 'skin' – with a focus on textures, patterns, surfaces and coverings. She is interested in adapting old typologies in new ways and working with traditional materials innovatively: she is currently trying to persuade Driade to produce 'Bentwood' chairs in aluminium.

Urquiola is well aware that we live in a pluralist society which demands that we be receptive to new ideas, and this is reflected within the selection of this book. Her work follows one route, but she has

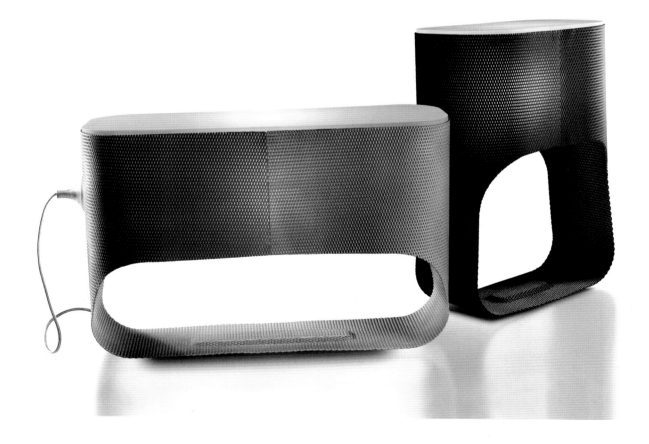

Lamp, Bague
Foscarini, 2003

respect for the avant-garde and the more conceptual designers, whose portfolios place as much emphasis on the story behind an object as on the object itself, and who are intent on creating physical comments on the design world. What she dismisses, however, are those products that give greater importance to aesthetics and narrative than to function. 'A personal style should be a good mix of concept, form and material use, none of which dominates the other. If you make things that are trendy, but not useful, you are not being a designer.' She recognizes a movement today towards the figurative, which places an emphasis on the more subjective and a greater consideration of our needs, desires and pleasures. It is something with which she herself is experimenting, but she is concerned that the 'look' could become the focus of attention rather than what she refers to as 'other defining elements'. As she writes in her foreword, 'The debate should never lead to a series of beautiful objects that don't function, whose value lessens when they are used rather than just admired. The figurative or decorative element should not be utilized as a short cut to avoid technological issues nor at the price of reduced practicality.' Semiotics and narrative are important to Urquiola, but only if there is a reason behind them; otherwise there

is a risk of slipping towards 'art' and away from industrial design.

Patricia is not alone in the concerns she raises. For the last decade, the design world has been in a state of flux, with no one design style or trend dominating. In one way this has been liberating and has opened up discussions on what design should be. The breaking down of barriers between disciplines, skills, roles and cultures has led to a greater experimentation and acceptance of individuality, the conceptual and low-tech, craft-orientated approaches. But in its wake there has emerged a movement away from industry and towards the ephemeral world of fashion, of attention-seeking design pieces and designers with rock-star status.

It is no coincidence that 'Art Basel', the world's biggest art show, is now gaining the reputation for being a venue for the exhibition of signature design objects. Extended to allow the world's leading design galleries to sell to art collectors who are beginning to love design as an investment as well as objects to be displayed, the show reflects the fact that design today is being bought like art, traded like art and increasingly produced like art. Hugues Magen of Magen H. Gallery, one of the seventeen New York art galleries invited to show at last year's 'Design Miami/Basel' (an offshoot of 'Art

Basel') says, 'There is this conversation about art and design meeting – form, function, art design; all those ideas are up for grabs at the moment.' The danger is that what was once conceived democratically and produced industrially is now being redefined by the way it is being consumed and collected, the market creating a demand for low-volume pieces that allow designers to make statements.

There is a genuine need to redress the balance. As far back as 1999, in the introduction to his edition of the yearbook, Jasper Morrison was talking about giving visual and perceptual order to a product combined with what he calls 'objectality' – the emotional response an item elicits. It's a quality that Patricia recognizes in his work: 'I love his work a lot, *capito*. He just does a little movement from the standard. It's very clever. He gets so much from such a little. It's the key to the elegance of his work.' When asked last year for which of his designs he would like to be remembered, Morrison replied, 'I'd prefer to be remembered by the sum of the work rather than any one thing. I don't think I'm trying to achieve anything spectacular, really – rather the opposite these days. I've decided to call it "Super Normal".' Morrison is much more vehement than Urquiola, but his sentiments echo her anxieties. 'Designers should be the guardians of the man-made

environment. Unfortunately, design [in general] has taken another course and has become a source of visual pollution. It's no longer a case of serving the consuming masses with objects that are easier to make industrially and that are better to live with. Now even big businesses are competing to make design as noticeable as possible by colour, shape or shock value. Things that are designed to attract attention are usually unsatisfactory.'

Choosing designs for the yearbook was not an easy undertaking for Urquiola. And her task was made all the more difficult as, for the first time, the book has been produced biennially and her selection had to cover two years, although the emphasis still remains on products from the last twelve months. The pages that follow give Patricia's personal selection from what she refers to as a pre-selection. Governed by a restricted time frame (this year's edition was completed on a very tight schedule) and the pre-qualifications of what was on offer, she has chosen not only 'the linear' – what we know and what is commercial – but also 'the arbi-trary': the wild card, the fresh and the innovative. Above all, she has sought to illustrate some of the themes, influences and researches that have given her pleasure. For any obvious omissions she apologizes: 'I've suffered, *capito*. Really suffered.'

Above:
Armchairs, Bloomy
Moroso, 2004

Opposite:
Armchairs, Fjord
Moroso, 2002

Right:
Sofa system, Tufty Time
B&B, 2005

Below:
Bed with integral
bedside table, Clip
Molteni, 2002

furniture

Chest of drawers, Eek Dresser
Piet Hein Eek
Steel
H: 100cm (39³⁄₈in)
W: 100cm (39³⁄₈in)
D: 50cm (19⁵⁄₈in)
Moooi, the Netherlands
www.moooi.com

Table, Brasilia
Fernando and Humberto Campana
Glass, metal
Various dimensions
Edra, Italy
www.edra.com

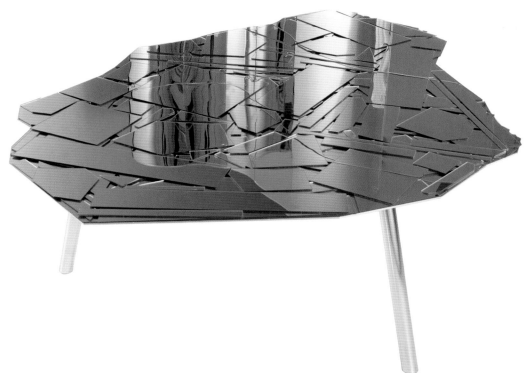

Cabinet, Multileg Cabinet
(Showtime Collection)
Jaime Hayón
Laquered wood
Single: H: 81cm (32in)
W: 52 cm (20½in)
L: 100–300cm (39 ⅜–118 ½in)
Double: H: 131 (51½in)
W: 52cm (20½in)
L: 100–300cm (39 ⅜–118 ½in)
Bd Ediciones de Diseño, Spain
www.bdbarcelona.com

Furniture, Micos: Monsterdog
Roberto Feo and Rosario Hurtado
El Ultimo Grito
Fibreglass
H: 40cm (15¾in)
W: 65cm (25⅝in)
L: 62cm (24⅜in)
Limited-batch production
El Ultimo Grito, Spain
www.elultimogrito.co.uk

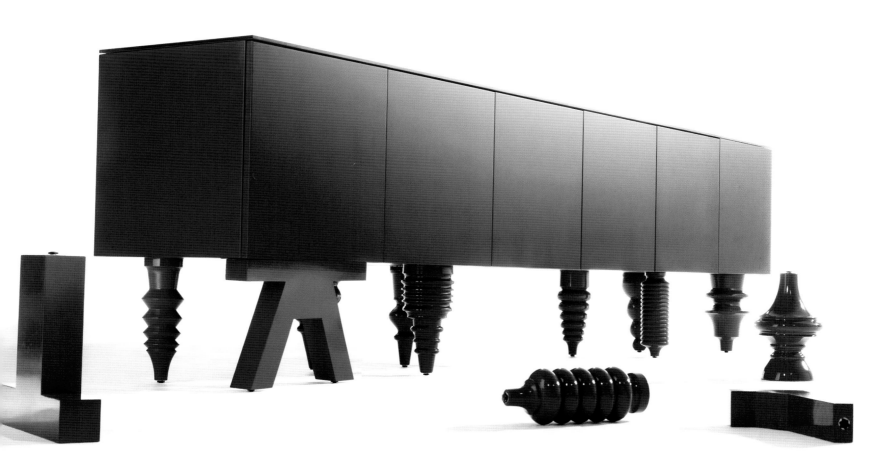

Armchair and pouf, Lotus
Jasper Morrison
Beech multiply, polyurethane foam,
die-cast polished aluminium
Chair: H: back 108cm (42½in)
seat: 39cm (15⅜)
W: 65cm (25⅝in)
D: 80cm (31½in)
Pouf: H: 39cm (15⅜in)
W: 44cm (17⅜in)
D: 37cm (14⅝in)
Limited-batch production
Cap Design, Italy
www.cappellini.it

Rocking Armchair, MT3
Ron Arad
Rotational moulded polyethylene
H: back: 78cm (31in)
seat: 50cm (19⅝ in)
W: 80cm (31½in)
D: 104cm (41in)
Driade, Italy
www.driade.it

Stool, Eskilstuna Stool
Graeme Findlay & Carmel McElroy
Steel, 100% wool
H: 40cm (15¾in)
W: 26cm (10¼in)
IKEA of Sweden, Sweden
www.ikea.com

Philippe Starck

The man or woman in the street would be hard-pressed to name more than a handful of architects – and even fewer designers. The exception is Philippe Starck. The *enfant terrible* of design, whose personality as well as his much-publicized prolific output has guaranteed him rock-star status: he is the most celebrated and commercially successful designer in the world.

Yet one senses today that maybe the tide is turning for Starck. Writing for *Metropolis* in March 2006, Philip Nobel refers to a 'Y2K crisis *chez* Starck, marking a moment of high ambition and higher insanity from a designer well known for both'. Quoting The Yoo condominiums, the design-for-everyone range of products for Target, and the plans for a spaceport in New Mexico, for Richard Branson, Nobel sees a watering down in Starck's recent work and a concurrent attempt by the designer to claw back some of the 'bad-boy street

cred that his success – and that of the designers he inspired – had helped to sap from his image'.

There is an element of truth in what Nobel writes, but it is not the whole story. In a lecture given by Starck at the Lisbon Experimenta Design Festival in 2005, it seems that the answer lies in a more deep-rooted sense of self-doubt, in the role of design and the designer in today's society: 'Design is useless. I'm ashamed of being a designer. Three or four years ago I gave my company to the people who work for me. I want to be less involved. I am no longer interested in architects or designers and I no longer want to talk about design ... I never stop producing but you might say I produce out of sheer idleness. I design in seconds, in spare moments. I'm a Faust-like character selling my soul to the devil: a hotel takes one and a half days, a boat an afternoon, and a product five minutes. The only solid thing in life is relationships between people and the vibration caused by a look. Quantum physics

expounds that nothing really exists – atoms, time and images. In that case, do we actually need anything? I continue to work for my ego, for money and fun, but there is no real reason.'

Of course this could be one of the outbursts to which Nobel refers but I think it would be more generous to interpret it as genuine questioning by a man used to being treated as a demigod in the small universe which is the fashion-led design world of today.

Armchair, Lago
Philippe Starck
Stiff polyurethane and anodized
aluminium, leather, lacquer
H: 80.5cm (31⅝in)
W: 60cm (23⅝in)
Limited-batch production
Driade, Italy
www.driade.it

Chair, Dickie
Anthony Kleinepier
Polypropylene laminated
with polyester, and nylon
with styrofoam filling
H: 120cm (47¼in)
W: 110cm (43¼in)
D: 105cm (41⅜in)
Moooi, The Netherlands
www.moooi.com

Sofa, Couch
Stefan Diez
Textile, polystyrene balls
H: 70cm (27½in)
W: 160cm (63in)
D: 75cm (29½in)
Elmar Floetotto, Germany
www.elmarfloetotto.de

Modular Sofa, Settanta
Enzo Berti
Wood frame filled with differentiated
density expanded polyurethane and
goose feather; fabric
H: 75cm (29½in)
L: 75cm (29½in)
D: 95cm (37⅜in) (each element)
Saba Italia, Italy
www.sabaitalia.it

Lounge chair, Prince Chair
Louise Campbell
Laser-cut metal, water-cut
EPDM, felt
H: 80cm (31½in)
W: 100cm (39⅜in)
D: 80cm (31½in)
Hay, Denmark
www.hay.dk

Armchair, Dora
Ludovica and Roberto Palomba
Polythene
H: 74cm (29⅛in)
W: 63cm (24¾in)
L: 70cm (27½in)
Zanotta, Italy
www.zanotta.it

Easy Chair, Link Easy Chair
Tom Dixon
Steel with white powder finish
H: 100cm (39⅜in)
W: 115cm (45¼in)
D: 70cm (27½in)
Tom Dixon, UK
www.tomdixon.net

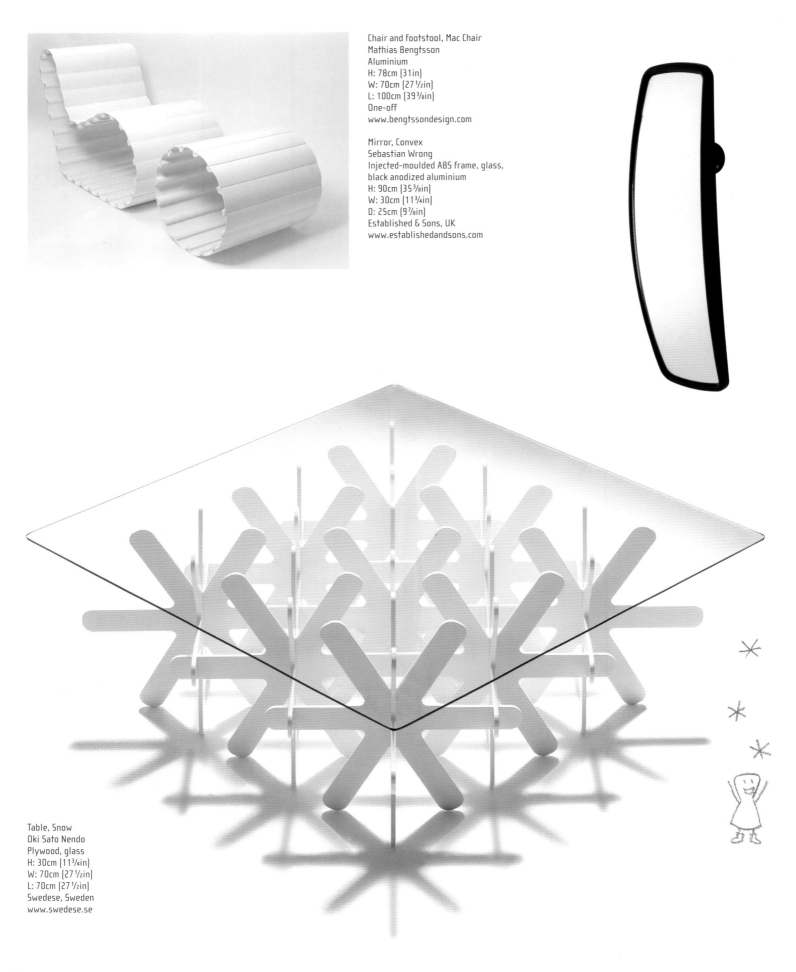

Chair and footstool, Mac Chair
Mathias Bengtsson
Aluminium
H: 78cm (31in)
W: 70cm (27 ½in)
L: 100cm (39⅜in)
One-off
www.bengtssondesign.com

Mirror, Convex
Sebastian Wrong
Injected-moulded ABS frame, glass,
black anodized aluminium
H: 90cm (35⅜in)
W: 30cm (11¾in)
D: 25cm (9⅞in)
Established & Sons, UK
www.establishedandsons.com

Table, Snow
Oki Sato Nendo
Plywood, glass
H: 30cm (11¾in)
W: 70cm (27½in)
L: 70cm (27½in)
Swedese, Sweden
www.swedese.se

Container cube, Optic
Patrick Jouin
PMMA
H: 40cm (15¾in)
W: 40cm (15¾in)
L: 40cm (15¾in)
Kartell, Italy
www.kartell.it

Nest of side tables, Polar
Oki Sato Nendo
Polarizing film, glass, steel
H: 36cm (14⅛in)
W: 50cm (19⅝in)
D: 50cm (19⅝in) (largest)
Prototype
Swedese, Sweden
www.swedese.se

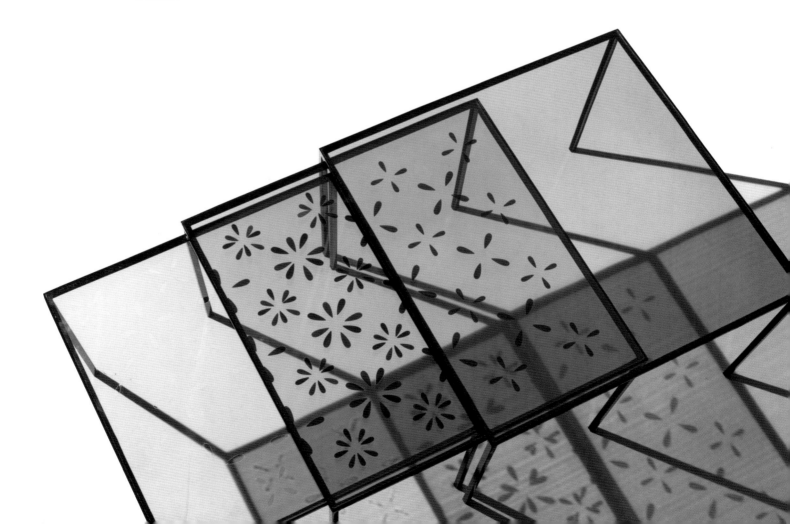

Chair, The Ultra Bellini
Mario Bellini
Nylon
H: 84cm (33in)
W: 44cm (17³⁄₈in)
D: 46cm (18¹⁄₈in)
Heller, USA
www.helleronline.com

Modular system,
Parts of a Rainbow
Christian Flindt
Acrylic satin
H: 86cm (34in)
W: 42cm (16¹⁄₂in)
L: 56cm (22in)
Prototype
Christian Flindt, Denmark
www.flintdesign.dk

Low lounge chair, Loom
Franco Poli
Frame: polished stainless
steel; cover: coach hide net;
feet: plastic
H: 99cm (39in)
W: 95cm (37in)
D: 73cm (28¾in)
Matteograssi, Italy
www.matteograssi.it

Armchair, Mr Bugati
François Azambourg
Fine metal sheet injected
with foam polyurethane
H: 70cm (27½in)
L: 60cm (23⅝in)
D: 62.5cm (24⅝in)
Cap Design, Italy
www.capellini.it

Chair, Supernatural
Ross Lovegrove
Polyamide, reinforced
with fibreglass
H: back: 81cm (32in)
seat: 46cm (18⅛in)
W: 53cm (20⅞in)
D: 51cm (20⅛in)
Moroso, Italy
www.moroso.it

Chair, A-Poc Gemini Ripple Chair
Ron Arad
Multi-layered techno-polymer
composite, steel; extruded fabric:
wool, cotton, down
H: back: 80cm (31in),
seat: 44cm (17⅜in)
W: 68cm (26¾in)
D: 59cm (23¼in)
Moroso, Italy
www.moroso.it

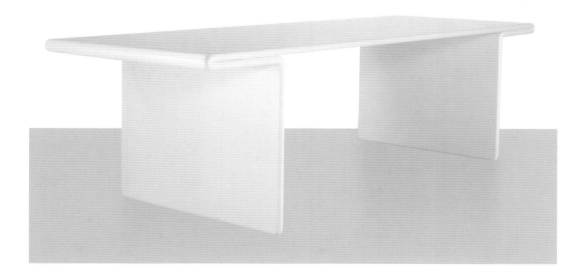

Table, Flex
Aziz Sariyer
MDF
H: 72cm (28⅜in)
W: 90cm (35⅜in)
L: 240cm (94½in)
Derin, Turkey
www.derindesign.com

Table, Balls
Bertjan Pot
Mixed wood
H: 75cm (29½in)
W: 100cm (39⅜in)
L: 240cm (94½in)
Moooi, The Netherlands
www.moooi.com

Ron Arad

Ron Arad's experimentation with form, process and material has put him at the forefront of contemporary design. Born in Tel Aviv in 1951, he arrived in London in 1973, enticed by the 'exoticism' of the city, which at that time was emerging from the peace-and-love generation of the 1960s into the anti-establishment, youth-fuelled, aggressive punk-rock years.

After graduating from the Architectural Association, Arad spent a brief period in a small architectural practice in Hampstead: 'It didn't take me long to realize that I didn't want to work for anyone. One lunchtime out of sheer boredom I walked out and I didn't go back. I went to a scrap yard behind the Roundhouse and started work on my first Rover car seat. That turned me into a furniture designer.'

In 1981 Arad founded his own company, One Off, producing individual pieces and ready-made items. The Rover 200 and the concrete stereo have become iconic, capturing as they did the *Zeitgeist* of the early 1980s with its rugged, streetwise DIY aesthetic and individualism set against a background of urban degeneration and disaffected youth. Arad, however, is very keen to point out that his work was not punk: it shared a certain rebelliousness and disregard for the rules but it was not angry. 'I'm a privileged person. I had nothing to rail against. I'm the product of an expensive education. I've always been interested in ready-made in art, in the poetic licence of Dadaism. I just expanded that into things I could

do. I was lucky to be in the right place at the right time and with the right sort of audience. If you think about it, London was a design desert at that time so I had to invent my own profession.'

Ron Arad Associates was formed in 1989 and One Off was incorporated into the company in 1993. Today the practice is involved in mass-manufactured products, gallery pieces and architecture. Whether he is designing a vase or an apartment block, Arad's work is always technologically advanced, experimental, inventive and challenging: he worked with plastic as a mass-manufactured material before anyone else, pioneered the technique of stereolithography, blew aluminium, explored and implemented the potential of implanting pixels in surfaces, animating volumes with still and moving images, and liberated the creative possibilities of Corian.

He is currently enthused by his collaboration with A-Poc, whose garments dress a limited edition of the Ripple chair, crossing the boundaries of fashion and industrial design. 'Issey Miyake and design engineer Dai Fujiwara have developed an incredible technique: they extrude garments. There is no sewing involved; it will eliminate the need for exploitative sweatshops, and the results are mass-produced but individual. The more sophisticated the machine gets, the less machine-like the product becomes. Working with them is a fantastic ongoing experience for me.'

Although Arad is better known for his industrial design work, that is about to change. A series of

high-profile architectural projects that have either just been completed (the floor in the much publicized Puerta America designer showcase hotel, and the Duomo, Arad's signature hotel in Rimini) or are currently being designed (the Swarovski hotel in Wattens [2004–], the Upperworld Hotel [2003–], which will sit at the top of the much-awaited restoration of London's Battersea Power Station, the Magis Headquarters in Treviso, the Holon Design Museum in Israel, an apartment block in Tel Aviv and a public sculpture in Corten steel in the centre of Jerusalem) prove that he has now bridged the gap from product designer to internationally acclaimed architect.

When asked what he thought made clients select him over the competition, he replied: 'Because we're fantastic. Because we have freedom of mind, we're not a good-taste option or a safe bet. It takes an adventurous person to choose us.'

Chair, Spring
Damian Williamson
Solid ash
H: 80cm (31½in)
W: 45cm (17¾in)
D: 51cm (20⅛in)
De Padova, Italy
www.depadova.it

Low chair, Wrapp
Marc Krusin
Steel rod, bent plywood, upholstery
H: 68cm (26¾in)
W: 79cm (31in)
D: 80cm (31½in)
Viccarbe Habitat, Spain
www.viccarbe.com

Bookcase, Random
Eva Paster and Michael Geldmacher
MDF
H: 216.3cm (85⅛in)
L: 81.6cm (32⅛in)
D: 25cm (9⅞in)
MDF Italia, Italy
www.mdfitalia.it

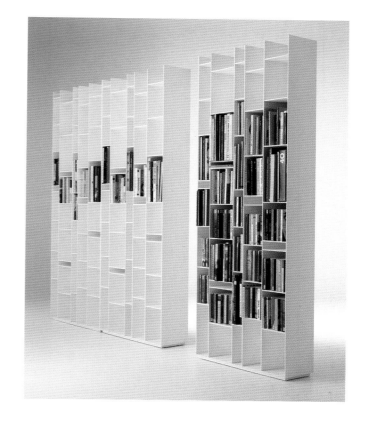

Office chair, Spoon Chair
Antonio Citterio and Toan Nguyen
Mass-coloured polypropylene
H: min. 44cm (17⅜in)
Seat: W: 50cm (19⅝in)
D: 43cm (17in)
Base: Diam: 70cm (27½in)
Kartell, Italy
www.kartell.it

Sofa range, Still
Foster & Partners
Cold-foamed metal, feather padding,
variable-density polyurethane,
chromed steel
H: 95 or 105cm (37³/₈ or 41³/₈in)
L: 220, 250 or 270cm (87, 98¹/₂
or 106¹/₄in)
D: 86cm (33⁷/₈in)
Molteni, Italy
www.molteni.it

Chair (opposite) and chaise longue,
Leaf and Leaf Longue
Alberto Lievore
Iron steel bars
Chair: H: 32.3 or 82cm (12³/₄ or 32¹/₄in)
W: 22 or 56cm (8⁵/₈ or 22in)
D: 20.5 or 52cm (8 or 20¹/₂in)
Chaise longue H: 31.2 or 78cm
(12¹/₄ or 30³/₄in), W: 29.2 or 73cm
(11¹/₂ or 28³/₄in), L: 53 or 61.2cm
(20⁷/₈ or 24in)
Arper, Italy
www.arper.it

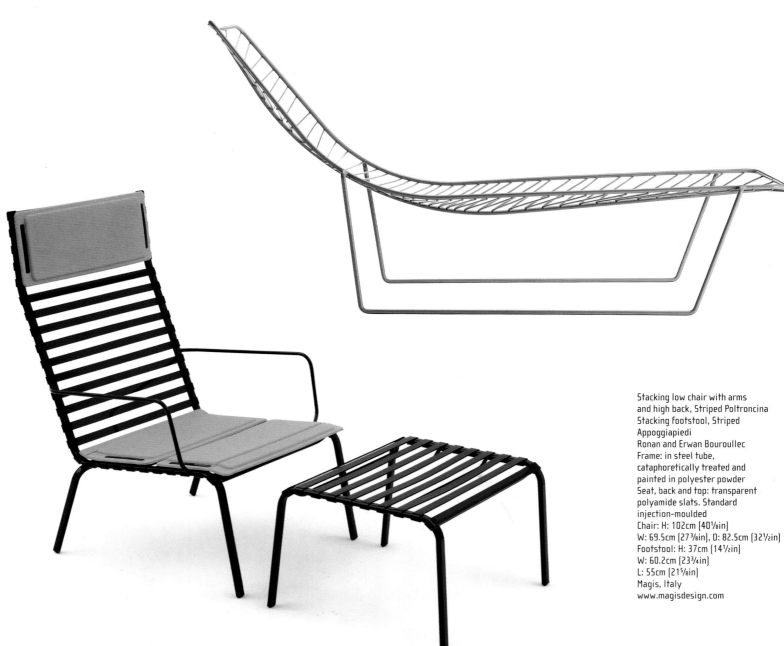

Stacking low chair with arms
and high back, Striped Poltroncina
Stacking footstool, Striped
Appoggiapiedi
Ronan and Erwan Bouroullec
Frame: in steel tube,
cataphoretically treated and
painted in polyester powder
Seat, back and top: transparent
polyamide slats. Standard
injection-moulded
Chair: H: 102cm (40¹/₈in)
W: 69.5cm (27³/₈in), D: 82.5cm (32¹/₂in)
Footstool: H: 37cm (14¹/₂in)
W: 60.2cm (23³/₄in)
L: 55cm (21⁵/₈in)
Magis, Italy
www.magisdesign.com

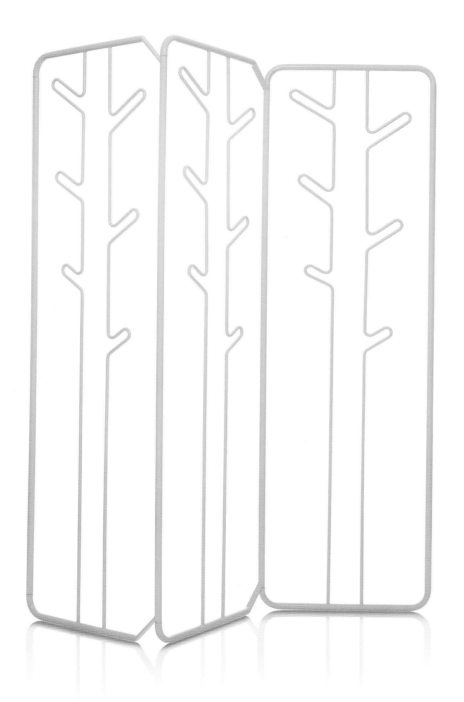

Coatstand/room divider, Boshetto
Fabio Biancaniello
Powder-coated steel tubing
H: 75cm (29½in)
L: 50 cm (19in)
Nanoo by Faser-Plast, Switzerland
www.nanoo.ch

Outdoor seating system, Meteor
Arik Levy
Polyethylene
Small: H: 30cm (11⅞in),
W: 50cm (19⅝in), L: 57cm (22½in)
Medium: H: 32cm (12⅝in),
W: 52cm (20 ½in), L: 87cm (34¼in)
Large: H: 34cm (13⅜in),
W: 60cm (23⅝in), L: 117cm (46in)
Serralunga, Italy
www.serralunga.com

Bench, Spun Bench
Mathias Bengtsson
Carbon Fibre
L: Section: 300cm (118in)
Diam: 50cm (19⅝in)
Limited-batch production
www.bengtssondesign.com

Maarten Baas

'If I wasn't a designer I would like to be a cook. What I don't like about design is the fact that people's opinion on art or design is often very rational. Experts, for example, want to see it in the context of their vision, or of history, and non-experts are often influenced by status or general taste. People often think they should "understand" design. What I miss in design is the personal, subjective aspect of the taste of food: just do you like it or not?'

Maarten Baas is the product of the concept-led school of design emanating from the Eindhoven Academy. Since graduating in 2002, his rise to fame has been exponential, thanks largely to his now iconic Smoke series and the support he received

from design entrepreneur Murray Moss, who commissioned Baas to produce blow-torched versions of design classics, which were put on show in his gallery in the 2004 exhibition 'Where There's Smoke'. 'I think the most I have done with my work is to tell some beautiful fairy tales. I have no idea where design is heading. I just make things that I like or consider to be interesting,' says Bass. He is an iconoclast who ploughs his own furrow, caring little about others' opinions.

The Clay Furniture does just what it says on the tin: 'Less the thought, the more the joy' was the explanation. 'We had industrial clay here for some other things and then I thought: hey, let's make furniture with that stuff. So I just started making it and it ended up as a collection!'

Chairs, Clay Furniture
Maarten Baas
Industrial clay, metal, coloured
lacquer, leather
Armchair: H: 82in (32in),
W: 41in (16⅛in), D: 46in (18⅛in)

Stacked dining chairs: H: 80cm (31½in),
W: 40cm (15¾in), D: 45cm (17¾in)
High-backed chair: H: 120cm (47in),
W: 40cm (15¾in), D: 45cm (17¾in)
Maarten Baas, The Netherlands
www.maartenbass.com

Artek, Bambu collection

Known for its innovation in wood technology, Artek has recently been focussing on the potential of bamboo, an often overlooked natural resource, which traditionally has been used mainly in woven basketwork or in flat board form. Bamboo is not a tree or a bush but is actually a 'super grass', which can be easily and quickly grown and harvested, making it ideal for sustainable design. Artek has concentrated its research into developing a technology to bend this material, which has a fibrous construction and a natural tensile strength; the result is an ingenious collection of leg, seat and surface components that allows a combination of table, chair and bench configurations.

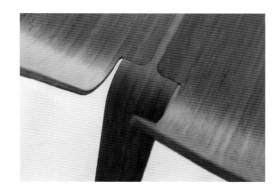

Chair and connected chairs,
Bambu Chair and Doublechair
Artek Studio led by Hendrik Tjaerby
Bamboo
Chair: H: 81cm (31⁷/₈in), W: 45cm
(17³/₄in), D: 41cm (16¹/₈in)
Doublechair: H: 81cm (31⁷/₈in),
W: 90cm (35³/₈in), D: 41cm (16¹/₈in)
Artek, Finland,
www.artek.fi

Stacking chair, Alma
Javier Mariscal
Polypropylene with glass fibre
added. Standard injection-moulded
H: 58cm (22⁷/₈in)
W: 39.2cm (15³/₈in)
L: 40cm (15³/₄in)
Prototype
Magis, Italy
www.magisdesign.com

Yves Béhar, Kada stool/table

Yves Béhar was brought up on a small island, Buyukada, offshore from Istanbul. It was there that he first noticed the simple low table used in Turkish coffee shops, its removable silver top allowing drinks to be served easily and ceremoniously to the guests. Béhar's twenty-first-century reworking is flat-packed and gains its strength when folded. A variety of tops from upholstered to glass allows the structure to be used in many ways. 'The numerous typologies and surfaces match the complexity of life with the simplicity of the product's construction and assembly. Ultimately, its use should be as simple as the Turkish coffee table.'

Table/stool, Kada
Yves Béhar
Wood, metal, fabric or plastic
material, or plexiglas/wood
H: 45cm (17¾in)
W: 35cm (13¾in)
L: 35cm (13¾in)
Danese Milano, Italy
www.danesemilano.com

Stacking armchair, Air-armchair
Jasper Morrison
Polypropylene with glass fibre added
H: 72.4cm (28½in)
W: 53.3cm (21in)
D: 50.5cm (19⅞in)
Magis, Italy
www.magisdesign.com

Stacking stool, Stool_One
Konstantin Grcic
Legs: anodized aluminium or
painted in polyester powder
Seat: die-cast aluminium painted
in polyester powder
H: 74cm (29in),
W: 53.4cm (21in)
L: 46cm (18⅛in)
Prototype
Magis, Italy
www.magisdesign.com

Stool, Paris Bar
Christophe Pillet
Aluminium die-casting,
mirror-polished aluminium
H: 65 or 82cm (25½ or 32¼in)
Diam: 50.3cm (19⅞in)
Driade, Italy
www.driade.it

Chair/trestle, Stool
Kensaku Oshiro
Plain or painted wood
H: 58cm (22⅞in)
W: 30cm (11¾in)
L: 50cm (19⅝in)
Prototype

Containers, Panier
Ronan and Erwan Bouroullec
Polycarbonate
H: 23cm (9in)
Diam: 61cm (24in)
Kartell, Italy
www.kartell.it

Furniture, Mico
Roberto Feo and Rosario Hurtado,
El Ultimo Grito
Rotational-moulded polyethylene
H: 40cm (15¾in)
W: 65cm (25½in)
L: 62cm (24⅜in)
Magis, Italy
www.magisdesign.com

Barber Osgerby, Pantone® stools

The concept of the stool is based on the Pantone® chip and the idea that colour is expressive of our emotions and individuality; the series consists of eight different colour sets with six shades within each, a total of forty-eight pantone colours. 'It brings the iconic chip to life in three dimensions. The chip is a symbol universally recognized by artists and designers.'

PANTONE®
16-0207 TPX

Stool, Pantone® Flight Stool
Edward Barber and Jay Osgerby
Birch plywood
H: 46cm (18⅛in)
W: 40cm (15¾in)
D: 39cm (15⅜in)
Limited-batch production
Isokon Plus, UK
www.isokonplus.com

Table, Cell
Eero Koivisto
Blizzard plastic
H: 36cm (14⅛in)
W: 100cm (39⅜in)
L: 100cm (39⅜in)
Limited-batch production
OFFECCT, Sweden
www.offecct.se

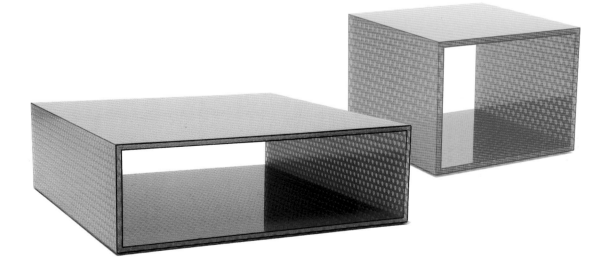

Chair, Shell chair
Edward Barber and Jay Osgerby
Plywood
H: 79cm (31in)
W: 43cm (17in)
D: 46cm (18⅛in)
Isokon Plus, UK
www.isokonplus.com

Entrance-hall and lobby furniture,
Upon floor
Stefan Diez
Powder-coated steel
H: 175cm (69in)
Diam: 50cm (19⅝in)
Schönbuch, Germany
www.schoenbuch.com

Stackable barstool, Miura
Konstantin Grcic
Reinforced polypropylene
H: 81cm (32in)
W: 47cm (18½in)
D: 40cm (15¾in)
Plank Collezioni, Italy
www.plank.it

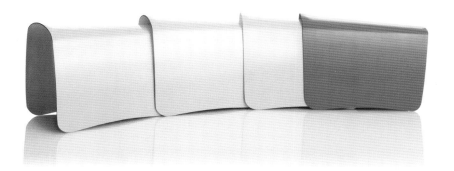

Bench
Emanuelle Jacques
Fibreglass
H: 48cm (19in)
W: 70cm (27½in)
D: 41cm (16⅛in)
Nanoo by Faser-Plast,
Switzerland
www.nanoo.ch

Indoor/outdoor furniture, Bent
Christophe de la Fontaine and
Stefan Diez
Sheet aluminium, powder-coated
H: 67cm (23⅜in)
W: 93cm (36⅝in)
D: 60cm (23⅝in)
Moroso, Italy
www.moroso.it

Stool, Onda
Jesús Gasca
Soft rubber, stainless steel
H: 83 or 93cm (33 or 37in)
W: 41cm (16⅛in)
Stua, Spain
www.stua.com

Chair, Smile (opposite, left)
Lievore, Altherr and Molina
Beech frame with oak seat
or upholstered seat
H: 78cm (31in)
W: 54.5cm (21½in)
D: 50cm (19⅝in)
Andreu World, Spain
www.andreuworld.com

Chair, Inglesina (opposite, right)
Ashley Hall and Matthew Kavanagh
Leather, plywood, chromed steel
tube and rod
H: 78cm (31in)
W: 48cm (18⅞in)
L: 62cm (24⅜in)
Gruppo Sintesi, Italy
www.sintesi2.it

Armchair/outdoor chair, Vela (left)
Hannes Wettstein
Steel, PVC
H: 78cm (31in)
W: 56cm (22in)
L: 62cm (24⅜in)
Accademia, Italy
www.accademiaitaly.com

Low table, Twisty
A + A Cooren
Corian and glass with
white silk-screened motif
H: 28cm (11in)
Diam: 90cm (35½in)
Prototype
Créa Diffusion, France
www.crea-diffusion.fr

Table, Lace Effect Table
Sharon Walsh and Andra Nelki
Polyester, vintage lace, linen
H: 41cm (16⅛in)
W: 55cm (21⅝in)
L: 55cm (21⅝in)
One-off
Sharon Walsh, UK

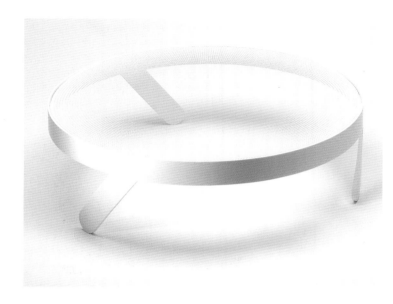

Coffee table, Blow up
Fernando and Humberto Campana
18/10 stainless steel, glass
H: 44cm (17⅜in)
Diam: 45cm (17¾in)
Alessi, Italy
www.alessi.com

Basket, coatstand and umbrella
stand, Bincan
Naoto Fukasawa
Plastic, steel
Basket: H: 38cm (15in),
Diam: 30cm (11¾in)
Coat stand: H: 170cm (67in),
W: 21.5cm (8½in),
Diam: 30cm (11¾in)
Umbrella stand: H: 84cm (33in),
Diam: 22cm (8⅝in)
Danese Milano, Italy
www.danesemilano.com

Coatstand, Tree
Mario Mazzer
Polyethylene
H: 175cm (69in)
Diam: 25cm (9⅞in)
Bonaldo, Italy
www.bonaldo.it

Stool, Miss Judith
Studio Demakersvan: Judith de Graauw
Kodiak and Highland leather
H: 45cm (17¾in)
W: 52cm (20½in)
D: 21cm (8¼in)
Limited-batch production
Montis, The Netherlands
www.montis.nl

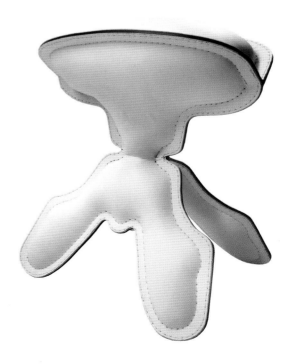

Folding chair, Honeycomb
Alberto Meda
Anodized aluminium,
technopolymer
H: 46cm (18⅛in)
W: 44cm (17⅜in)
D: 57cm (22½in)
Kartell, Italy
www.kartell.it

Chair, Spline
Norway Says
Metal rod with a rubber
surface treatment
H: 76cm (30in)
W: 54cm (21¼in)
D: 48cm (18⅞in)
Limited-batch production
OFFECCT, Sweden
www.offecct.se

Screen, Fiore
Fabrizio Bertero, Andrea Panto
and Simona Marzoli
Laser-cut steel
H: 197cm (78in)
W: 265cm (104in)
D: 41cm (16⅛in)
Limited edition
Zanotta, Italy
www.zanotta.it

Table support, Coral
Formstelle Studio
Plywood
H: 72cm (28⅜in)
W: 70cm (27½in)
D: 48cm (18⅞in)
Limited-batch production
Novum 3, Germany
www.novum3.de

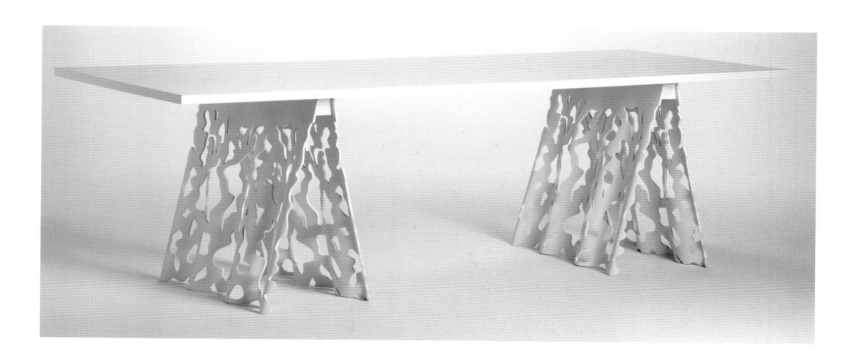

Dining table, Decorating Dining
Lucy Merchant
Highly polished to mirror finish
stainless steel
H: 77cm (30in)
W: 66cm (26in)
L: 178cm (70in)
Prototype
Lucy Merchant UK
www.lucymerchant.com

Table, Campo d'Oro
Paolo Pallucco and Mireille Rivier
Steel and laminate
H: 73cm (28¾in)
W: 140cm (55⅛in)
L: 140cm (55⅛in)
De Padova, Italy
www.depadova.it

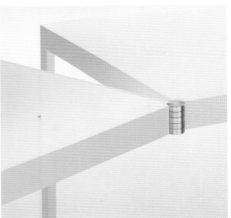

Table, Table First
Stefano Giovannoni
Polyamide, MDF, wood
or HPL laminate
H: 71cm (28in)
L: 130cm (51in)
W: 80cm (31in)
Magis, Italy
www.magisdesign.com

Stacking chair, Family First
Stefano Giovannoni
Air-moulded polyamide
H: 78cm (31in)
W: 50cm (19⅝in)
L: 52cm (20½in)
Prototype
Magis, Italy
www.magisdesign.com

Chair, Clipt
Jeff Miller
Polycarbonate shell,
steel base
H: 75.5cm (29¾in)
W: 47.5cm (18¾in)
D: 52.5cm (20⅝in)
Baleri Italia, Italy
www.baleri-italia.com

Table, Synapsis
Jean-Marie Massaud
Plated steel, Pral
H: 73.2cm (28⅞in)
W: 90 or 120cm (35⅜ or 47in)
L: 350cm (137¾in)
Porro, Italy
www.porro.com

Trunk/container, El Baúl
Javier Mariscal
Rotational-moulded
polyethylene
H: 56cm (22in)
W: 61 cm (24in)
L: 91cm (35⅞in)
Magis, Italy
www.magisdesign.com

Sofa, Ishi
Naoto Fukasawa
Wood, polyurethane
foam, leather
H: 47.5cm (18¾in)
W: 150cm (59in)
D: 118cm (46in)
Mass production
Driade, Italy
www.driade.com

Seat, Truffle
Jean-Marie Massaud
Thermoplastic material, metal
H: 44cm (17³⁄₈in)
Diam: 110cm (43³⁄₈in)
Porro, Italy
www.porro.com

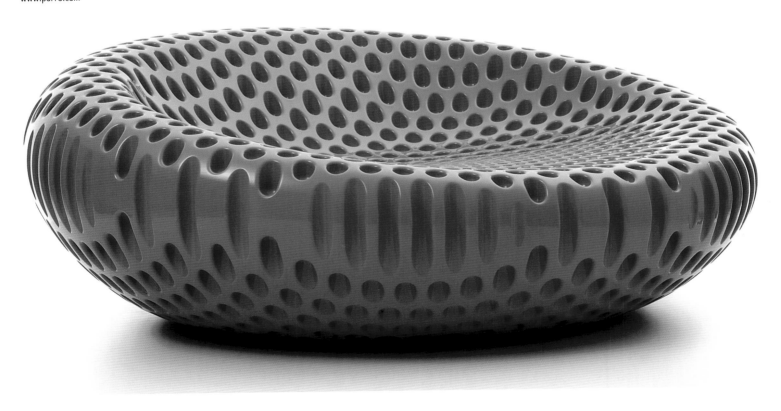

Stool, Lilla
Patrick Norguet
Polyurethane foam, fabric
or leather
H: 55cm (21⅝in)
W: 36cm (14⅛in)
D: 38cm (15cm)
Production on order
Artifort, The Netherlands
www.artifort.com

Garden furniture, Uno
Phillip Eckhoff, Matthias Kübler,
Isabel Schurgacz
Injection-moulded plastic
H: 120cm (47¼in)
W: 50cm (19⅝in)
D: 35cm (13¾in)
Gandia Blasco, Spain
www.gandiablasco.com

Seating system, Arne
Antonio Citterio
Tubular steel, flexible cold-shaped
polyurethane, die-cast and
extruded brushed aluminium
L: 320cm (126in), W: 151cm (59½in)
L: 315cm (124in), W: 166cm (63⅜in)
L: 309cm (121⅝in), W: 166cm (63⅜in)
B&B Italia, Italy
www.bebitalia.it

Sofa/loveseat, Odin
Konstantin Grcic
Base: synthetic resin
Upholstery: polyurethane
with padding
Cover: fabric or leather
H: 73.5cm (29in)
W: 69cm (27⅛in)
L: 160cm (63in)
Classicon, Germany
www.classicon.com

Sofa, Print
Marcel Wanders
Wooden structure, polyurethane
and polyester fibre
H: 85cm (33½in)
W: 270cm (106¼in)
D: 96cm (38in)
Moroso, Italy
www.moroso.it

Sofa, Polder
Hella Jongerius
Wooden frame with belt
upholster, polyurethane chips
and microfibres, upholstered
polyurethane foam and
polyester wool
Cushion buttons: horn, olive
wood, bamboo and mother
of pearl
H: back: 78cm (31in),
seat: 41cm (16in)
L: 333cm (131in)
D: 100cm (39⅜in)
Vitra, Switzerland
www.vitra.com

Floor cushion, Bovist
Hella Jongerius
Textiles, linen, polyester,
polypropylene balls,
granulate
H: 38cm (15in)
Diam: 54cm (21¼in)
Vitra, Switzerland
www.vitra.com

Hella Jongerius

It was with some surprise that the design press received the news that Hella Jongerius was going mainstream with the quartet of vases she designed for IKEA's 2006 PS collection (see page 188), but hey, they were just ceramics so that didn't count. There is now no denying the fact. The Polder sofa and Bovist pouf she has designed for the latest Vitra Home collection means she is now well and truly taking on the big boys.

Trained at the Eindhoven Academy and with an early alliance with Droog, she has forged an industrial-meets-craft style that has gained her a leading position as one of the most innovative and interesting product designers working today.

On the eve of the 2003 solo show at London's Design Museum, Alice Rawsthorne, then director, was quoted as saying 'One of the most important themes in contemporary design is to imbue industrially produced objects with the character that people have traditionally loved in handcrafted pieces, and Hella is at the forefront.'

To capitalize on her success, then, is it any wonder that she has now made the conscious decision to collaborate with large global manufacturers, an industry she has been critical about in the past? She states that one of her reasons to 'hit the high street' is to influence mass-market design from within to show that the avant-garde can translate commercially. 'I resist the neutral perfection and

conformity we generally associate with industrial production. But instead of rejecting industry, I want to search for solutions from within it.'

Polder is inspired by techniques such as tailoring, collage and embroidery, which Jongerius has combined with a toned-down, contemporary design language and the use of coordinated colours and textures. With quiet irony, the name is a reference to the tracts of land reclaimed from the sea in the Netherlands, the lines of the sofa recreating the horizontality of this kind of terrain. Special attention has been paid to the conspicuously large decorative buttons, made of exotic natural materials such as horn, bamboo and mother of pearl.

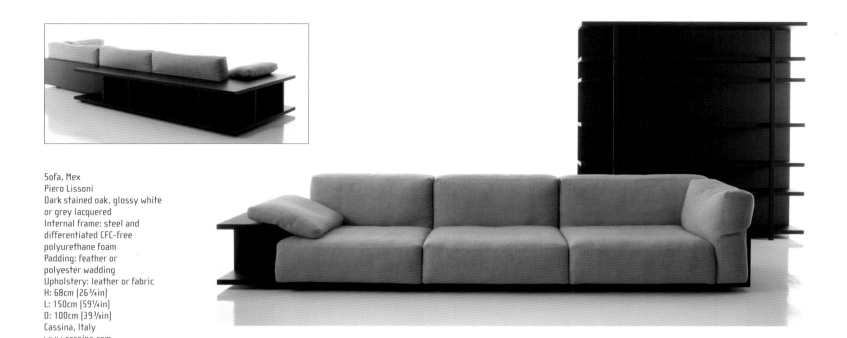

Sofa, Mex
Piero Lissoni
Dark stained oak, glossy white
or grey lacquered
Internal frame: steel and
differentiated CFC-free
polyurethane foam
Padding: feather or
polyester wadding
Upholstery: leather or fabric
H: 68cm (26¾in)
L: 150cm (59¼in)
D: 100cm (39⅜in)
Cassina, Italy
www.cassina.com

Padded chairs, TT range
Alfredo Häberli
Polyurethane, CFC-free
polyurethane foam, steel tubing,
die-cast aluminium; polyester,
new wool or leather
Various dimensions
Alias, Italy
www.aliasdesign.it

Bookcase, Brera
Lievore, Altherr and Molina
Honeycombe, chipboard, MDF
H: various
W: 160, 210 or 300cm
(63, 83, 118½in)
D: 33cm (13in)
EmmeB Industria Mobili, Italy
www.emmebidesign.com

Divan, Belt
Vico Magistretti
Wood, steel, polyurethane
H: 42 or 80cm (16½ or 31½in)
W: 162, 228 or 294cm
(64, 90 or 116in)
D: 90cm (35⅜in)
De Padova, Italy
www.depadova.it

Coffee table, Flatpack Furniture
Maarten Baas
Wood, glass
H: 40cm (15¾in)
Diam: 110cm (43⅜in)
Maarten Baas, the Netherlands
www.maartenbaas.com

Side table with tray;
Animal Thing, Pig Tray
Front Design
Glass fibre
H: 65cm (25½in)
Diam (tray): 43cm (16⅞in)
Moooi, The Netherlands
www.moooi.nl

Sofa, Late Sofa
Ronan and Erwan Bouroullec
Cloth or leather for the
upholstery, aluminium
H: 88.5cm (35in)
W: 254cm (100in)
D: 83cm (33in)
Vitra, Switzerland
www.vitra.com

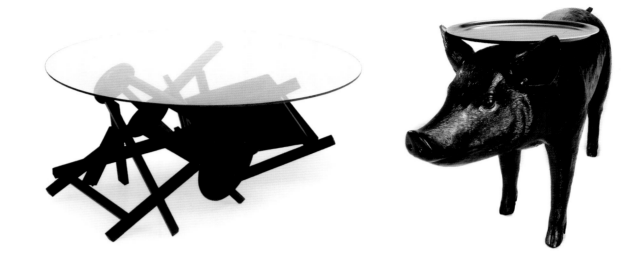

Sofa, Basket
Ronan and Erwan Bouroullec
Metal, fabric/leather
H: 82cm (32¼in)
W: 230cm (90½in)
D: 90cm (35⅜in)
Cap Design, Italy
www.cappellini.it

Sofa, Cloud
Naoto Fukasawa
H: 75cm (29½in)
W: 300cm (118½in)
D: 216.5cm (85¼in)
B&B Italia, Italy
www.bebitalia.it

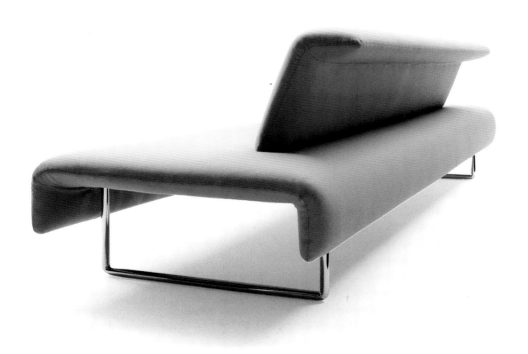

Patrick Jouin, One Shot stool

Jouin's latest piece produced by the Rapid Prototyping technique known as Selective Laser Sintering is the One Shot stool. A kinetic, integrated, functional design piece, it is produced in a single, non-assembled unit. The stool emerges from the machine in its final form, with all axles, screws, springs and hinges concealed by the structure of the stool itself. By virtue of gravity combined with a simple, elegant, soft-turning twist, the array of rods transforms in one flowing movement to a small, useful and aesthetically strong seat.

Foldable stool, One Shot
Patrick Jouin
Polyamide
Folded L: 66cm (26in)
Diam: 11cm (4³⁄₈in)
Open: H: 40.5cm (16in)
Diam: 32cm (12½in)
Materialise.MGX, Belgium
www.materialise-mgx.com

Cantilevered chair/stool,
La Robe (Family)
Julia Läufer and Marcus Keichel
Stainless steel, warp knitted fabric
H: 83cm (32⁵⁄₈ in)
W: 57cm (22½in)
D: 60cm (23⁵⁄₈ in)
One-off

Patrick Jouin, Solid furniture

Patrick Jouin's Solid furniture is a whole new way of looking at furniture manufacture. Using the process of stereolithography (a technique originally invented by Chuck Hall in 1986 for rapid proto-typing), he is literally growing furniture in his studio. Patrick Jouin says, 'What's interesting is that we are able to grow objects structurally in the same sort of way that nature works.'

Taking as his inspiration forms from the living world – cells, grass reeds, crystals – shapes are created using a CAD program. Jouin then uses a 3D layering system to build up successive sheets of 0.5mm (¼in) plastic, each individually cut and photo-chemically hardened by laser before adding the next layer. Eventually the plan is to create bespoke furniture, with customers selecting from a range of colours, forms, textures and materials, which are then developed in the lab into whatever item is desired and delivered a week later.

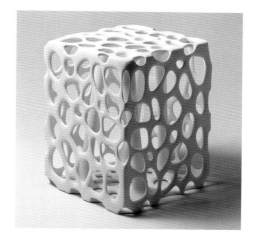

Stool, Solid S1
Patrick Jouin
Stereolithography using epoxy resin
H: 43.8cm (17¼ in)
W: 33.7cm (13¼ in)
L: 33.7cm (13¼ in)
Limited edition
Patrick Jouin Studio, France
www.patrickjouin.com

Chair, Solid C2
Patrick Jouin
Stereolithography using epoxy resin
H: 77cm (30¼in)
W: 42cm (16½in)
D: 53cm (20⅞in)
Materialise.MGX, Belgium
www.materialise-mgx.com

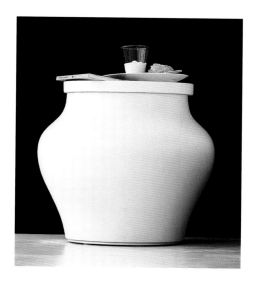

Night table, Dream
Marcel Wanders
Matt and glossy lacquered colours, wenge, and ovangkol recomposed
H: 41.9cm (16½in)
Diam: 45cm (17¾in)
Poliform, Italy
www.poliform.com

Foersom and Hiort-Lorenzen, Imprint chair

It's not very often that a completely new material is developed, and certainly not one that is so obvious it has been totally overlooked. In this age of increasing ecological responsibility, why is it only now that someone has come up with the idea of farming the basic structure of all plant cell walls, cellulose, on an industrial scale?

Readily available, recyclable and environmentally sound, Cellupress has been developed by Foersom and Hiort-Lorenzen in collaboration with Lammhults and Dan-Web (a material manufacturer) and has been used in the former's new Imprint chair. Soft cellulose mats are treated with non-carcinogenic glue, fed into a machine and compressed using very high temperatures and a lot of pressure. The result is as strong and hard as wood but with the smoothness and versatility of a plastic. To give Cellupress varying textures and colours, spruce, coconut and oak have been added, enhancing the material from within.

The chair is produced by imprinting the human form onto square panels. As Cellupress is absorbant, the surface is then treated with a non-toxic lacquer. The legs have been kept as minimal as possible to concentrate on the shape of the fibre shell.

Chair, Imprint
Johannes Foersom and
Peter Hiort-Lorenzen
Pressed wooden fibres
H: 78cm (31in)
W: 53 or 55cm (20⁷⁄₈ or 21⁵⁄₈in)
D: 58 or 62cm (22⁷⁄₈ or 24³⁄₈in)
Lammhults, Sweden
www.lammults.se

Chair, Amìla
Naoto Fukasawa
Metal, polyurethane
H: 81cm (32in)
W: 43cm (16⁷⁄₈in)
D: 51cm (20¹⁄₈in)
Danese Milano, Italy
www.danesemilano.com

Armchair, Hola
Hannes Wettstein
Chair: steel, CFC-free
polyurethane foam,
black plastic material;
covers: fabric and
soft leather
H: 79cm (31in)
L: 52cm (20½in)
D: 55cm (21⅝in)
Cassina, Italy
www.cassina.com

Chair, Soft Boing
Tokujin Yoshioka
Aramid honeycomb paper
H: back: 79cm (31in),
seat: 43cm (16⅞in)
W: 58cm (22⅞in)
D: 78cm (31in)
Limited-batch production
Driade, Italy
www.driade.com

Chair, Ottochair OT/5
Antonio Citterio and
Toan Nguyen
Chromed tubular steel,
cast aluminium,
fabric or leather
H: 78cm (31in)
W: 50cm (19⅝in)
D: 59cm (23¼ in)
B&B Italia, Italy
www.bebitalia.it

Seating System, Facett
Ronan and Erwan Bouroullec
Foam, quilting, stitching
Various dimensions
Ligne Roset, France
www.ligne-roset.com

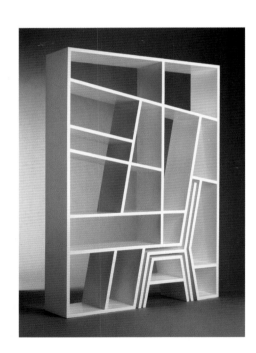

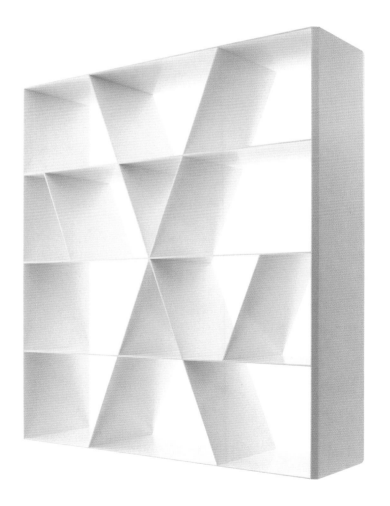

Shelves/desk, Shelflife Desk
Charles Trevelyan
Lacquered LDF (low-density MDF)
H: 180cm (71in)
L: 196cm (77in)
D: 36cm (14⅛in)
Viable, UK
www.viablelondon.com

Bookcase, Shelf X
Naoto Fukasawa
White acrylic
H: 145.5cm (57¼in)
L: 131cm (51½in)
D: 37cm (14⅝in)
B&B Italia, Italy
www.bebitalia.it

Kram and Weisshaar, Breeding tables

Kram and Weisshaar's backgrounds are very different. Reed Kram was born in Ohio and specializes in media design, having cut his teeth designing video games prior to co-founding the Aesthetics and Computation Group led by John Maeda. Weisshaar's background is in product design and he was assistant to Konstantin Grcic before setting up his own company. Their studio was founded in 2002, yet the duo work in different cities (Stockholm and Munich, respectively), and collaborate by means of frequent meetings and the latest communication technology. They like the idea of a two-centre office as it allows them to build up more contacts, experiences and potential for growth.

The Breeding Table project was the result of the disillusionment they experienced during the Milan Furniture Fair in 2002. With their varying backgrounds and expertise in IT and advanced technological manufacturing methods, they were disappointed that not enough companies were harnessing these two very important developments

to produce something completely different. Once back, they started to look at different computer programs, experimenting with algorithmic modelling to produce subtle repeats of a table shape. The couple are often quoted as describing the computer and software they use (Rhino – much beloved of industrial designers) as their 'digital sweatshop'. From the information it's given, it churns out hundreds of unimaginable designs, which the duo then select, construct on laser-cutting and steel-bending machines, and pass on to highly trained technicians to finish by hand. Their intention is to prove that mass-manufacture and craftsmanship can co-exist, which they believe is the only way that European design industry will be able to compete with the threat from the production lines of the Far East.

Kram and Weisshaar are not alone in their desire to combine personalization with mass production. Ron Arad and Ross Lovegrove have long worked with stereolithography, Patrick Jouin is also experimenting in this field with yet another technique (*see* page 61), Marcel Wander's Snotty vases for

Rosenthal trace the trajectory of a sneeze digitally and re-create it industrially, and each one of Gaetano Pesce's Nobody's Perfect range of resin furniture for Zerodisegno is unique although mass-produced.

What differentiates Kram and Weisshaar, however, is their total examination of the design process from concept to realization – from how to get the most out of a digitized cutting machine normally used merely for repetitions, to exploring the potential of such a manufacturing technique and the effect it will have on the future design economy. We can expect to hear a lot more from this resourceful pair.

Software dynamically generating
tables, Breeding Table
Reed Kram and Clemens Weisshaar
Steel ST37, laser-cut and
powder-coated
Various dimensions
Moroso, Italy
www.moroso.it

Parasol, Shadylace
Chris Kabel
Lace, parasol frame
H: 205cm (81in)
Diam: 235cm (93in)
Symo, Belgium for Droogdesign,
The Netherlands
www.symo.be
www.droogdesign.nl

Low table, Oak Table
Jasper Morrison
Natural solid oak
H: 28cm (11in)
L: 121cm (48in)
D: 38.5cm (15⅛in)
Limited-batch production
Cap Design, Italy
www.capellini.it

Series of benches, a stool
and a storage table, Log
Naoto Fukasawa
Oak veneer
Benches: H: 40cm (15¾in),
L: 100 or 150cm (39⅜ or 59in),
W: 50cm (19⅝in)
Stool: H: 40cm (15¾in),
L: 50cm (19⅝in), W: 50cm (19⅝in)
Storage table: H: 48.5cm (19in),
D: 40cm (15¾in), W: 50cm (19⅝in)
Customer orders
Swedese Möbler, Sweden
www.swedese.se

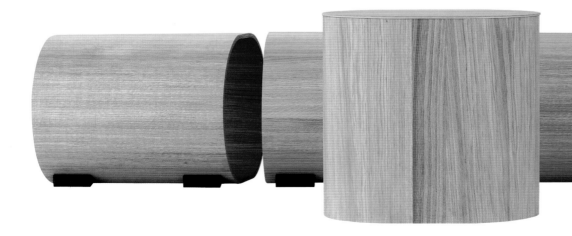

Table, DT3
Inoda+Sveje
Traditional 2D-bent plywood
(white oak), bent stainless-steel tubes
H: 73cm (28¾in)
Diam: 100cm (39⅜in)
Prototype
Italy

Jasper Morrison, Crate bedside table

Although coining the term 'Super Normal' for his philosophy of design – meaning that objects should measure up to the reality of everyday life and, whether created anonymously or conceived aesthetically, should be perceived through their use – Jasper Morrison nevertheless came in for a lot of stick for his Crate bedside table for Established & Sons, which smacks of the found object rather than a piece of autonomous design. It is in fact a wine crate, which has been only minutely altered (the joints were enforced and pine was substituted for Douglas Fir) from the original, which stood by Morrison's bed for three years.

Morrison's justifications are threefold: firstly, design is becoming a serious source of visual pollution, pandering to an image-hungry media – the box seeks to find a egoless solution. Secondly, apart from plagiarism, there should be no limits to how a designer achieves a design, so long as it's useful, good value and doesn't contaminate; the box was by his bed, he came across something useful and made it available. Finally, a lack of debate in the industry has resulted in manufacturers producing anything with a name attached to it. Morrison is angered that design is becoming a 'cheap-trick supplier to the media' – the box attempts to balance the situation.

Table/storage, The Crate
Jasper Morrison
Douglas Fir
H: 50cm (19⅝in)
L: 37.5cm (14¾in)
D: 17.5cm (6⅞in)
Established & Sons, UK
www.establishedandsons.com

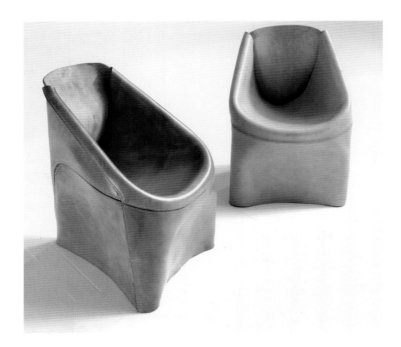

Chests, stands, cabinets,
Flatpack Antiques
Gudrun Lilja Gunnlaugsdóttir
Plastic-coated plywood
Various dimensions
Studiobility, Iceland
www.bility.is

Armchair, Raw
Tomek Rygalik
Steel, satined stainless steel,
expanded polyurethane,
leather rawhide
H: 75cm (29½in)
W: 60cm (23⅝in)
D: 65cm (25⅝in)
Prototype
Moroso, Italy
www.moroso.it
www.tomekrygalik.com

Bentwood chair
Estefanía Fernández and
Alfredo Sandoval
Beech-wood strips
H: 81cm (31⅞in)
W: 53cm (20⅞in)
D: 40cm (15¾in)
Prototype
Scuola Politecnica di Design;
Gebrüder Thonet Vienna, Italy
www.scuoladesign.com
www.thonet-vienna.com

Chair, 544
Piero Lissoni
Wood, steel tubing, polyester mesh,
fine wire
H: 76cm (30in)
W: 55cm (21⅝in)
D: 53cm (20⅞in)
Gebrüder Thonet, Germany
www.thonet.de

Furniture, Zoe
Lievore, Altherr and Molina
Twill cotton, expanded polystyrene
H: 72cm (28⅜in)
L: 120cm (47¼in)
D: 108cm (43in)
Verzelloni, Italy
www.verzelloni.it

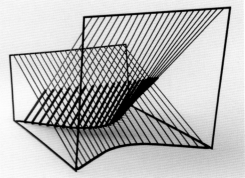

Lounge chair, 5MM
Mikael Mantila
8mm and 5mm steel rod,
rubber tube,
painted finish
H: 60cm (23⅝in)
W: 60cm (23⅝in)
L: 85cm (33⅜in)
Prototype

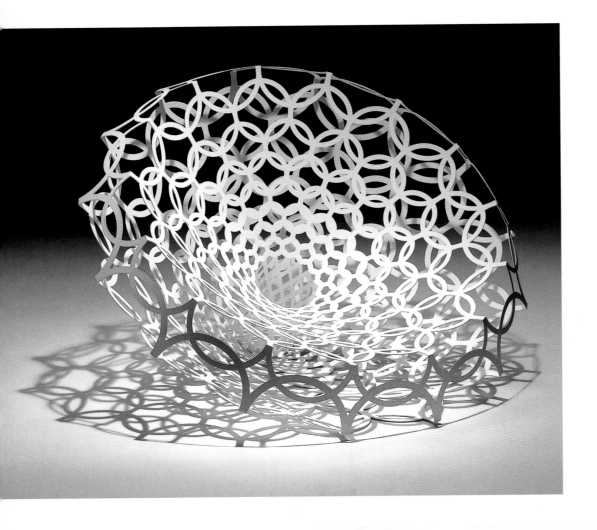

Outdoor chair, Veryround
Louise Campbell
2mm-thick steel sheet frame
cut by three-dimensional laser,
varnished for outdoor use
H: 69cm (27⅛in)
W: 105.5cm (41½in)
D: 83cm (33in)
Zanotta, Italy
www.zanotta.it

Stool, Stone
Marcel Wanders
Polycarbonate
H: 45cm (17¾in)
Diam: 30cm (11¾in)
Kartell, Italy
www.kartell.it

Dining chair, Flower Chair
Marcel Wanders
Chromed steel
H: 64cm (25¼in)
W: 78cm (31in)
D: 71cm (28in)
Moooi, The Netherlands
www.moooi.com

Table, Oval Table
Tord Boontje
Top: tempered glass printed in
black with a steel base powder-
coated in high-gloss black
H: 76.2cm (30in)
W: 125.7cm (49½in)
L: 219.7cm (86½in)
Moroso, Italy
www.moroso.com

71

Sofa, Aspen
Jean-Marie Massaud
Steel, nylon, fabric
or leather
H: 72cm (28⅜in)
L: 260cm (102⅜in)
D: 90cm (35½in)
Cassina, Italy
www.cassina.it

Room divider, Screen
Front Design
Plexi
H: 150cm (59in)
W: 120cm (47¼in)
D: 30cm (11¾in)
Front, Sweden
www.frontdesign.se

Stool/occasional table, Glaçon
Lee West
Ceramic
H: 35cm (13¾in)
W: 35cm (13¾in)
L: 35cm (13¾in)
Ligne Roset, France
www.ligne-roset.com

Stool, Royal T
Philippe Starck
Polyethylene
H: 75cm (29½in)
D: 35cm (13¾in)
Kartell, Italy
www.kartell.it

Five leather elements,
Pools & Pouf!
Robert Stadler
Leather
Various dimensions
Galerie Dominique
Fiat, France
www.radidesigners.com

Chair, Vera
Teemu Järvi
Chromed or powder-coated
12mm steel-rod frame; plywood;
wood or upholstery
H: 77cm (30³⁄₈in)
W: 52cm (20½in)
L: 58cm (22⁷⁄₈in)
Hkt-Korhonen, Finland
www.hkt-korhonen.fi

Chair, Wieki
Bertjan Pot
Steel, foam, polyester, rope
H: 78cm (31in)
L: 50cm (19⁵⁄₈in)
D: 51cm (20¹⁄₈in)
Pallucco, Italy
www.pallucco.com

Dining table and chairs,
New Antique
Marcel Wanders
Lacquered wood
Chair: H back: 78cm (31in),
H seat: 45cm (17¾in),
W: 55cm (21⅝in), D: 45cm (17¾in)
Dining table: H: 75cm (29½in),
W: 200cm (79in), D: 90cm (35in)
Cappellini, Italy
www.cappellini.it

Bookshelf, Line
Yedidia Blonder
Aluminium, stainless steel
Various dimensions
Yedidia Blonder, Israel
yblonder@bezeqint.net

Rocking horse/footstool, Candore
Giovanni Levanti
Steel, polyurethane, various fabrics
H: 160cm (63in), seat: 75cm (29½in)
W: 8–35cm (3⅛–13¾in)
L: 88cm (35in)
Campeggi, Italy
www.campeggisrl.it

Armchair, PL 100
Piero Lissoni
Steel, plastic, leather
H: 63cm (24¾in)
W: 92cm (36¼in)
D: 90cm (35in)
Fritz Hansen, Denmark
www.fritzhansen.com

Tables, PL703 and PL 803
Piero Lissoni
Powder-coated steel,
polyurethane lacquer, MDF beech
PL703: H: 45cm (17¾in), L: 49cm
(19¼in), W: 49cm (19¼in)
PL803: H: 55cm (21⅝in), L: 49cm
(19¼in), W: 49cm (19¼in)
Fritz Hansen, Denmark
www.fritzhansen.com

Table, Pipeline
Piero Lissoni
Extra-light float glass,
borosilicate glass, steel
H: 72cm (28⅜in)
L: 220cm (87in)
D: 90cm (35⅜in)
Glas Italia, Italy
www.glasitalia.com

Coffee table, Bentz JM202
Jeff Miller
Aluminium cast base, glass top
JM202: H: 35cm (13¾in)
Diam: 115cm (45in)
Baleri Italia, Italy
www.baleri-italia.com

Low table, Split
Piero Lissoni
Lacquered float glass,
banded extra-light glass
H: 30cm (11¾in)
Diam: 80 or 120cm (31½ or 47¼in)
Glas Italia, Italy
www.glasitalia.com

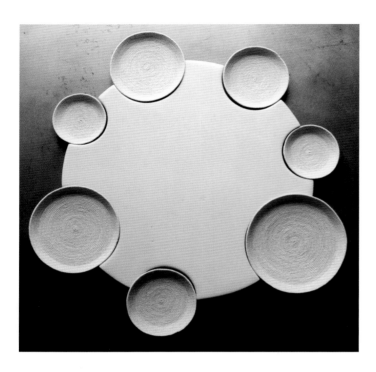

Table, Flore
Marta Laudani and
Marco Romanelli
Ceramic and raffia
Diam: 50cm (19⅝in)
Limited-batch production
SardegnaLab, Italy
www.sardegnalab.com
www.imagomundi.com

Divider, Roc
Ronan and Erwan Bouroullec
Cardboard panel with
textile surface
Various dimensions
Prototype
Vitra, London
www.vitra.com

Shelves, T5
Martin Szekely
Laquered aluminium type 4G,
Nextel paint
H: 259cm (102in)
W: 377cm (148½in)
D: 46cm (18⅛in)
Limited-batch production
Galerie Kreo, France
www.galeriekreo.com

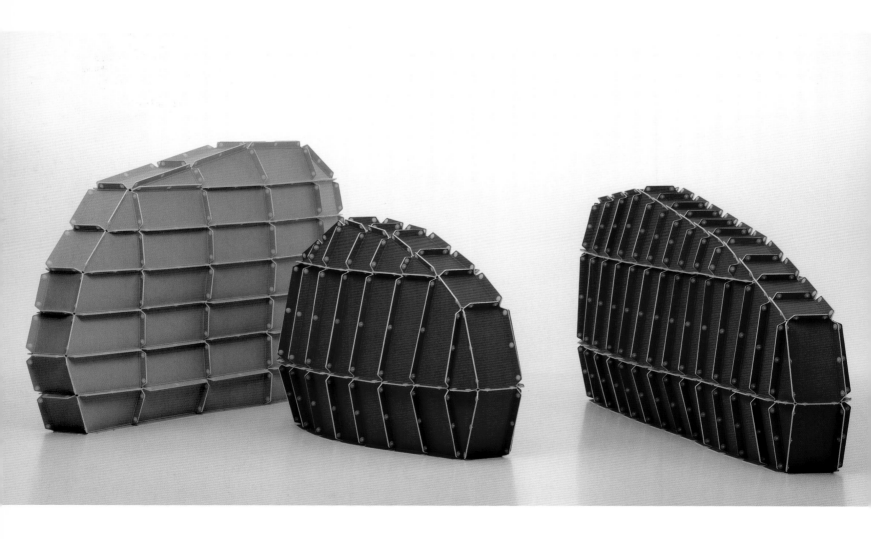

Armchair, Boing
Tokujin Yoshioka
Chromed steel, shell in foam
with metal inserts and fixed
cover in pleated leather
H: 87cm (34in)
W: 56cm (22in)
D: 78cm (31in)
Driade, Italy
www.driade.com

Armchair, South Beach
Cristophe Pillet
Birch wood,
satin-finish nickel-plated steel
H: 139cm (55in)
W: 110cm (43³⁄₈in)
D: 68cm (26³⁄₄in)
Tacchini, Italy
www.tacchini.it

Urban/transit furniture,
Los Bancos Svizos
Alfredo Häberli
25mm diam. steel tube,
2mm perforated steel sheet,
hot-dip galvanized or cataphoretic
coating finished with bronze-
coloured polyester resin
Various dimensions
Bd Ediciones de Diseño, Spain
www.bdbarcelona.com

Kiki van Eijk, Soft Cabinet

With a birch plywood construction and drawers and handles in ceramic, the Soft Cabinet has the deceptive appearance of padded leather. It is handmade with the handles glazed with real gold. The drawers are conceived modularly; any object, from a desk to a sideboard, can therefore be made on request.

Cupboard, Soft Cabinet
Kiki van Eijk
Ceramics, birch plywood
H: 225 cm (89in)
W: 156 cm (61³⁄₈in)
D: 32 cm (12¹⁄₂in)
Limited-batch production
Kiki van Eijk, The Netherlands
www.kikiworld.nl

Chair, Ole
Ludovica and Roberto Palomba
Plywood
H: back: 76cm (29⁷⁄₈in)
seat: 45cm (17³⁄₄in)
D: 51cm (20¹⁄₈in)
W: 55cm (21⁵⁄₈in)
Crassevig, Italy
www.crassevig.com

Dining table, Raw Table
Garth Roberts
Wood (oak) and varnished steel
H: 95cm (37³⁄₈in)
W: 74cm (29¹⁄₈in)
L: 240 or 279cm (94¹⁄₂ or 110in)
Limited-batch production
Zanotta, Italy
www.zanotta.it

Benches, Twig
Russell and John Pinch
Coppiced hazel
H: 46cm (18¹⁄₈in)
L: 150cm (59in)
D: 46cm (18¹⁄₈in)
Limited-batch production
Pinch, UK
www.pinchdesign.com

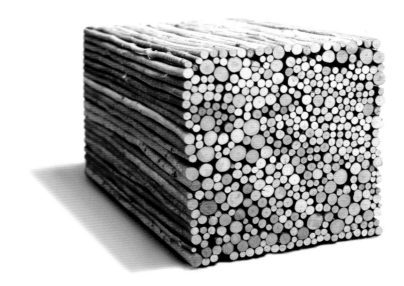

Table, Chab-table (far left)
Oki Sato Nendo
Wood, steel
H: 35.5 or 53cm (14 or 20⅞in)
D: 60cm (23⅝in)
De Padova, Italy
www.depadova.it

Low table, Tumb Tumb
Lorenzo Damiani
Polyurethane
H: 56cm (22in)
Diam: 45cm (17¾in)
Campeggi, Italy
www.campeggisrl.it

Bookcase, Pierced Bookcase
Andrea Branzi
Stainless steel, steel cables,
wicker, crystal
H: 195cm (77in)
W: 270cm (106in)
D: 50cm (19⅝in)
Design Gallery Milano, Italy
www.designgallerymilano.com

Table, Cinderella
Jeroen Verhoeven
Finnish birch
H: 81cm (32in)
W: 100 cm (39⅜in)
L: 134cm (53in)
Limited-batch production
Demakersvan, The Netherlands
www.demakersvan.com

Table, Baghdad
Ezri Tarazi
Industrial steel
H: 35 or 74cm (13¾ or 29⅛in)
W: 160cm (63in)
L: 200cm (79in)
H: 35 or 74cm (13¾ or 29⅛in)
W: 100cm (39⅜in)
L: 300cm (118½in)
Edra, Italy
www.edra.com

Table, Tivoli
Massimo Barbierato
Wood, iron
H: 77cm (30⅜in)
W: 70cm (27½in)
L: 270cm (106¼in)
One-off
2730project, Italy
www.2730project.com

Table, Bend-in
Arik Levy
Ash, Cataletto walnut treated with
oil, laminated top, heartwood
H: 73cm (28¾in)
W: 90 or 140cm (35½ or 55⅛in)
L: 200cm (79in)
Desalto, Italy
www.desalto.it

Storage cupboard, Riddled
Steven Holl
Walnut veneer with vegetable-oil
finish; aluminium anodized
H: 70cm (27½in)
L: 200cm (79in)
D: 50cm (19⅝in)
Horm, Italy
www.horm.it

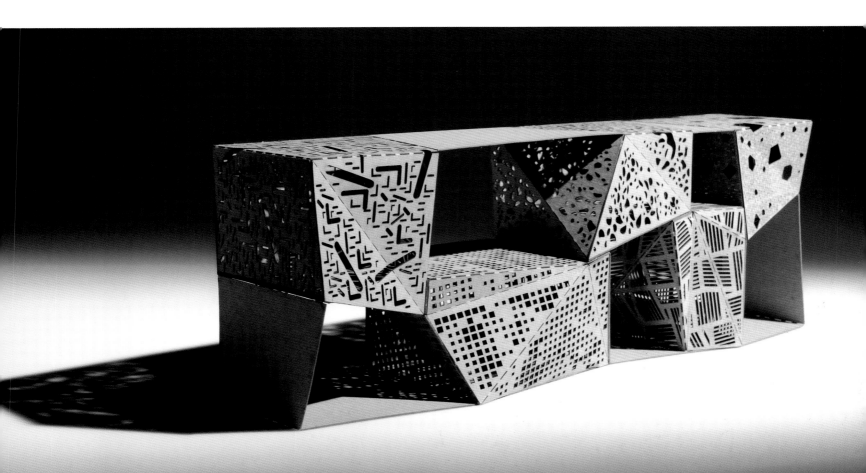

Coffee table, Jonker
Christopher Coombes and
Cristiana Giopato
10mm curved float glass
H: 31cm (12¼in)
W: 89cm (35in)
D: 89cm (35in)
Fiam Italia, Italy
www.fiamitalia.it

Coatrack, Ribbon
Voon Wong and Benson Saw
Painted mild steel
H: 180cm (71in)
Diam: 35cm (13¾in)
Limited-batch production
Voon Wong and
Benson Saw, UK
www.voon-benson.com

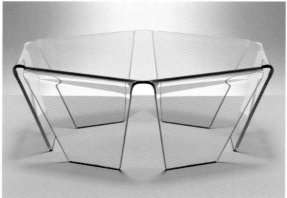

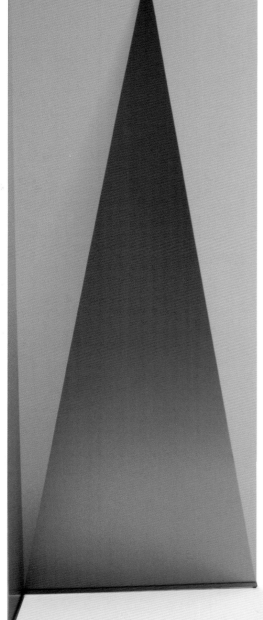

Screen, Shiki
Setsu and Shinobu Ito
Aluminium sheet
H: 161cm (63⅜in)
W: 85cm (33½in)
D: 42.5cm (16¾in)
De Padova, Italy
www.depadova.it

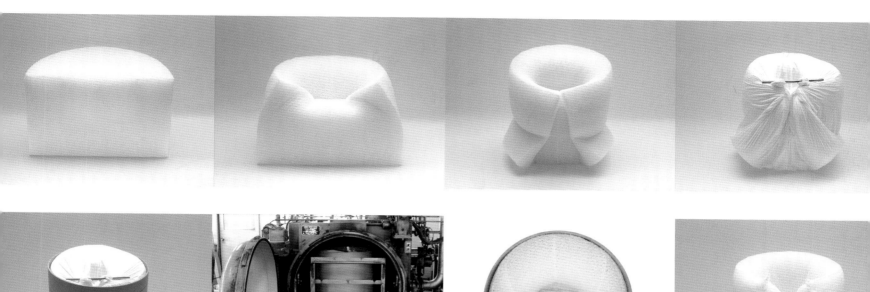

Chair, Pane Chair
Tokujin Yoshioka
Polyester elastomer
H: 80cm (31⅜in)
W: 90cm (35⅜in)
D: 90cm (35⅜in)
Tokujin Yoshioka Design, Japan
www.tokujin.com

Chair, IKEA PS Ellan
Chris Martin
Wood-fibre composite
H: 9.5cm (3¾in)
W: 68.5 (30in)
L: 79cm (31in)
IKEA of Sweden, Sweden
www.ikea.com

Chair, Mummy
Peter Traag
Bent beech, polyurethane foam,
polyester elastic ribbon
H: 84cm (33in)
W: 46cm (18⅛in)
D: 52cm (20½in)
Edra SpA, Italy
www.edra.com

Tokujin Yoshioka, Pane chair

Tokujin Yoshioka studied design with Issey Miyake and Shiro Kuramata. He established his own studio in 2000, which dedicates much time and space to experimentation. The Pane chair was inspired by an article that he read in *National Geographic* on the wonders of fibres and textiles. He was taken by the soft-looking but incredibly strong structures that are heavy but not solid, airy but not hard, and have a great capacity to absorb forces. The name 'Pane' comes from the Italian word for bread, which Yoshioka selected because the process behind the development of the chair closely resembled cookery.

Pane is made from a translucent spongy material called polyester elastomer. A half cylinder is wrapped in a sheet, the ends tied and knotted together. It is then baked in an oven to fix the shape. Once the sheet is removed, the ends of the arms take on the shape of the twisted textile. Yoshioka prefers autonomous and accidental forms, which go beyond the consciousness and overcome personal likes or dislikes.

Futon, Kuutio
Sirpa Fourastié and Susan Elo
100% cotton, upholstery, paper yarn
cotton fabric
Dimensions when folded as a cube:
H: 70cm (27½in)
W: 70cm (27½in)
L: 70cm (27½in)
Woodnotes, Finland
www.woodnotes.fi

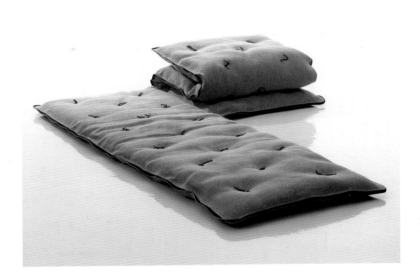

Mat/stool, A Needle and Thread
Gregory Lacoua
Beech ply, closed-cell rubber foam,
technical jersey in polyester fibre
Mat: W: 120cm (47¼in),
L: 120cm (47¼in)
Stool: W: 60cm (23⅝in), L: 60cm
(23⅝in)
Prototype
France

Armchair, Kiss Me Goodbye
Tokujin Yoshioka
Polycarbonate
H: 85cm (33½in)
W: 56cm (22in)
Limited-batch production
Driade, Italy
www.driade.it

Bench, Loop
Christophe Pillet
Polyethylene
H: 40cm (15¾in)
L: 180cm (71in)
D: 50cm (19⅝in)
Serralunga, Italy
www.serralunga.com

Perches and stools, Sway
Thelermont Hupton
Polyethylene, polyurethane
H: 34, 50 or 66.5cm
(13⅜, 19⅝ or 26⅛in)
Diam: 21cm (8¼in)
Thelermont Hupton, UK
www.thelermonthupton.com

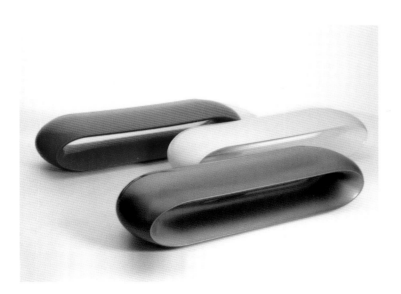

products

Tent, Hypno EX
Cam Brensinger and Suzanne Turell
Inflatable airbeam structure,
sail-cloth laminate, silicone-
impregnated nylon and Hypalon
H: 92.7cm (36½in)
W: 134.6cm (53in)
L: 213.4cm (84in)
Nemo Equipment, USA
www.nemoequipment.com

Compact tyre-repair system,
Air Repair System
Gerhard Reichert
Metal, plastic
H: 10.5cm (4⅛in)
Diam: 18.3cm (7⅛in)
Terra-S, Germany
www.terra-s.com

Hypno EX tent

This lightweight tent has dispensed with
traditional tent poles and pegs in favour of two
pairs of 'airbeams' connected internally to a single
inflation and deflation valve. The tent takes only
forty-five seconds to blow up by means of an
integrated pump, and ten seconds to deflate. The
shell consists of a silicone-infused fabric, which is
good for ventilation but at the same time protects
the interior from moisture.

Pillows, Handles
Hella Jongerius
Corduroy, leather, wool
Vitra International,
Switzerland
www.vitra.com

Cabin bag with briefcase,
UPPTÄCKA
Knut and Marianne Hagberg
EVA foam, 100% polyester,
aluminium
Cabin bag: H: 20cm (7⅞in),
W: 38cm (15in),
L: 52cm (20½in)
Briefcase: H: 20cm (7⅞in),
W: 33cm (13in),
D: 10cm (3⅞in)
IKEA of Sweden, Sweden
www.ikea.com

93

Knife sharpener, Chantry Modem
Sam Hecht, Industrial Facility
Pressure die-casting, butchers steel
W: 8.5cm (3¼in)
H: 8.5cm (3¼in)
L: 20cm (7⅞in)
Harrison Fisher, England
www.harrison-fisher.co.uk

Trivet, Bronzeknuckle
Ashley Hall and Matthew Kavanagh
Bronze
L: 22cm (8⅝in)
W: 18cm (7⅛in)
H: 2cm (¾in)
Limited-batch production
DDA Global, USA
www.diplomatdesign.com

Dish rack, Gradient Dish Rack
Leon Ransmeier and
Gwendolyn Floyd
Polypropylene
L: 42cm (16½in)
W: 29.7cm (11⅝in)
H: 7.3cm (2⅞in)
Prototype
Ransmeier and Floyd,
The Netherlands

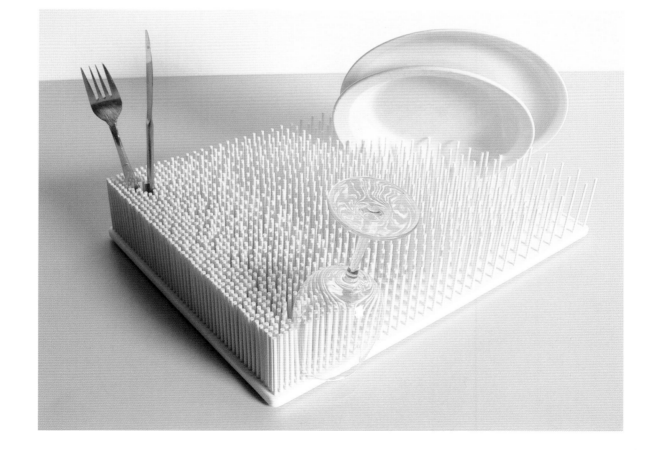

Sponge, Kitchen Sponge
Carlo Contin
Expanded polyurethane
H: 4.5cm (1¾in)
W: 8cm (3⅛in)
L: 13cm (5⅛in)
Coop, Italy
www.coop.it

Clothes peg, Washing Peg
Giulio Iacchetti
Polypropylene
W: 3.5cm (1⅜in)
L: 6.4cm (2½n)
D: 0.7cm (¼in)
Coop, Italy
www.coop.it

Grill pan, Vege
Mikko Laakkonen
Cast iron
Diam: 27cm (10⅝in)
Selki-Asema, Finland
www.selki-asema.fi

Cookware, Pots and Pans series
Jasper Morrison
18/10 stainless steel
Various dimensions
Alessi, Italy
www.alessi.com

Milk saucepan, BluBlu
Lorenzo Damiani
Pyrex
H: 15cm (6in)
W: 25cm (9⅞in)
D: 10cm (4in)
Prototype
Lorenzo Damiani Studio, Italy
lorenzo.damiani@tin.it

Stainless-steel cookware, Cookware
Jan Hoekstra
Stainless steel
Diam: 16cm (6¼in), 18cm (7⅛in),
20cm (7⅞in) and 24cm (9½in)
RoyalVKB, The Netherlands
www.royalvkb.com

Baking tools, Tools You Bake
Sebastian Summa and
Hrafnkell Birgisson
Aluminium
H: 8–15cm (3¹⁄₈in–6in)
Diam: 15–26cm (6in–10¼in)
Hugo Bräuer and Artificial, Germany
www.hugo-braeuer.de
www.artificial.de

Summa and Birgisson, Tools You Bake

Sebastian Summa and Hrafnkell Birgisson have developed baking moulds from industrial forms. Summa trained as a blacksmith and discovered the family-run tooling and metal-spinning company, Hugo Bräuer Metallwaren, while developing an assignment for his product design studio at the University of Applied Sciences in Potsdam. Inspired by the hundreds of dusty wooden moulds lining the shelves of the workshop, he recreated the story of an old industrial company in a modern domestic idiom.

The series of cake tins, Collatz, Wiesner, Etoga, Sturickow, Stubbak amd Bessy, are all named after former clients of the company, which produces round metal objects for industry and the military through a process in which sheets of metal – steel, copper and aluminium – are pressed around rotating moulds with a hand tool. Summa and Birgisson like to work with small companies as they are more flexible, and it is much easier to work with the craftsman firsthand, offering the possibility of producing small quantities to custom orders. 'These moulds fit perfectly into the concept of doing pastries that are dramatic yet cost-effective – you don't need the extra labour usually required to make special shapes.'

Konstantin Grcic, Krups espresso machine XP5000

The French kitchen appliances manufacturer SEB now has nine prestigious brands under its umbrella (including Krups, Rowenta and Moulinex), each with clearly demarcated territories and styles. In order to keep the resultant strategic advantage of this wide coverage of all market segments, SEB has begun a programme to reinforce the individual and well-defined brands by approaching various internationally renowned designers to add their personal touch to a range of new products.

Konstantin Grcic was asked to design a coffee maker for Krups as SEB believed that his style suited the 'precise, professional and structured' qualities of the marque. This was followed by a Panini maker, a state-of-the-art blender, which was launched during New York's International Contemporary Furniture Fair, 2006, and, a few months later, a cutting-edge food mixer.

Before his collaboration with Krups, Grcic was new to home-appliance design. He took as his starting point the brand itself and devised a series of visual clues or 'formal codes' that he recognized as being quintessential to the company – professional-quality goods, durability, user-friendliness – and then translated these into particular shapes, materials and control designs, applying them to all the different products in the series so that there is something 'Krups-like' in each appliance, no matter how different its function is from another.

With a simple, straightforward, central articulation of all functioning parts, including the handle to the coffee chamber and the steaming mechanism for milk, the espresso machine immediately looks professional. Following Grcic's belief that a machine should be able to be operated without instructions, there are no graphics on the appliance apart from a small arrow under the Krups logo, which indicates to the user where the handle should be placed but which also forces attention on the branding every time he or she makes a coffee.

Espresso machine, XP5000
Konstantin Grcic
PP, stainless steel
H: 31.7cm (12½in)
W: 21.5cm (8½in)
D: 41 cm (16⅛in)
Krups (Group SEB), France
www.krups.com

Packaging, Milk Cartons
Yael Mer
Laminated paper, cardboard
H: 20cm (7⅞in)
W: 7.5cm (3in)
D: 7.5cm (3in)
Prototype
www.yaelmer.com

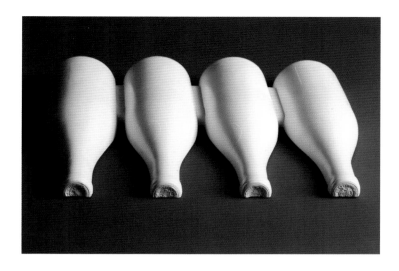

Wine rack, Pile
Harry Allen
Cast polyester resin
W: 29.2cm (11½in)
L: 35.6cm (14in)
D: 5cm (2in)
Areaware, USA
www.areaware.com

Package design, Pizza Box
Lotta Fagerholm
Paper, paperyarn, seal lacquer
H: 5cm (2in)
Diam: 30cm (11¾in)
University of Art and Design,
Helsinki, Sweden

Bottle, Ípsei
Edward Barber and Jay Osgerby
Polyethylene Terephthalate (PET)
H: 18.5cm (7¼in)
Diam: 5.5cm (2⅛in)
Coca Cola, Belgium
www.coca-cola.com

Wine thermometer, Vignon
Jakob Wagner
Steel and plastic
Diam: 7.5cm (3in)
Menu A/S, Denmark
www.menu.as

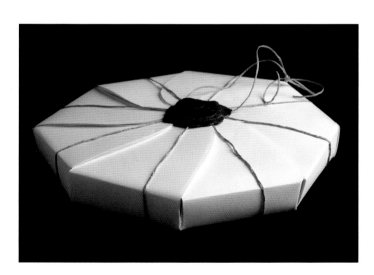

Mobile kitchen unit, Erika
Storno: Henrik Drecker,
Katharina Ploog, Sven Ulber,
Davide Siciliano
Wall panel FU (birch plywood),
laminate red high gloss; FU (birch
plywood), black FU (birch plywood),

stainless steel, sheet steel,
plated panels
H: 29–95cm (11³⁄₈–37³⁄₈in)
W: 29–62cm (11³⁄₈–24³⁄₈in)
D: Max 31cm (12¹⁄₄in)
Nils Holger Moormann, Germany
www.moormann.de

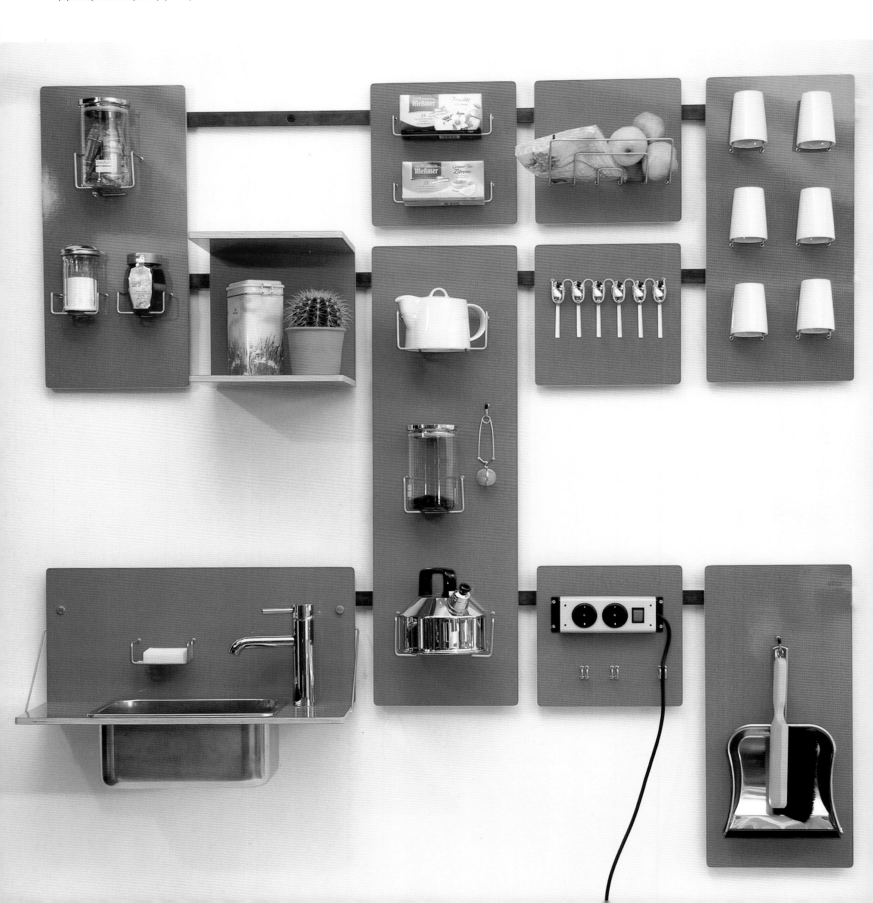

Kitchenette
Jan Jannes Dijkstra
Coated metal wire
Various dimensions
Prototype
Machinefabriek van der Hoorn,
The Netherlands
www.vanderhoorn.nl

Oven, 3D
Bruno Lizotte and Lynn Turnbull
Stainless steel and glass
H: 88.5cm (35¼in)
W: 59.5cm (23⅜in)
D: 54.5cm (21½in)
Electrolux, Sweden
www.electrolux.com

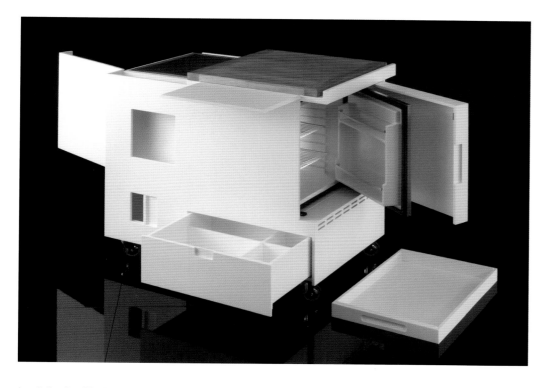

Joe Colombo, Minikitchen

Joe Colombo's life was cut tragically short when he died of heart failure at the very young age of forty-one. Yet during the brief period he had available to him he managed to leave behind many proofs of his creativity, some of which are as topical today as they were when they were designed in the 1960s. By experimenting with new materials and using advanced technologies, he developed 'machines for living' – multifunction mobile units, such as the Minikitchen, which was re-edited by Boffi in 2006.

Joe (Cesare) Colombo's early interest was science but he switched to fine art while still at secondary school and went on to study painting and sculpture at the Accademia di Belle Arti in Brera, Milan. He came to design relatively late when he took over his father's electrical conductor factory where he experimented with the latest production processes and newly developed plastics such as fibreglass, ABS, PVC and polyethylene, developing his adventuristic and futuristic trademark style in a series of memorable products that included the Universale, the first chair to be moulded from a single material, to the all-in-one living systems for the future. Concentrating on industrial design, he believed that good domestic design should be available to everyone and reflect new living patterns.

The Minikitchen was a small, one-piece unit on wheels, which contained all the electrical appliances and necessary features needed to cook and accommodate six people – in just half a cubic metre. Boffi's new version recreates the essential in a structure made of Marino multilayered panels and 12mm (½in) white Corian.

Monobloc kitchen on castors, Minikitchen
Joe Colombo
Marino multi-layered panels, Corian, induction hotplate in ceramic glass, 50 lt. mini-refrigerator, teak chopping board
H: 102cm (40in)
L: 102cm (40in)
W: 65cm (25⅝in)
Boffi, Italy
www.boffi.com

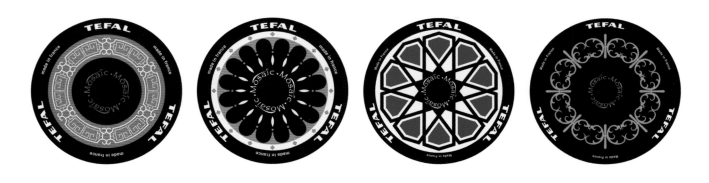

Cooking equipment, Mosaic Range
Doshi Levien
Cast aluminium, flu-turned
aluminium, terracotta, Bakelite
Tagine: H: 25cm (9⁷⁄₈in),
Diam: 35cm (13³⁄₄in)
Karhai: H: 16cm (6¹⁄₄in),
Diam: 37.5cm (14³⁄₄in)
Wok: H: 19cm (7¹⁄₂in), L: 50cm
(19⁵⁄₈in), Diam: 32cm (12⁵⁄₈in)
Fajita: H: 7.5cm (3in), W: 21cm
(8³⁄₈in), L: 45cm (17³⁄₄in)
Tefal, France
www.tefal.com

Table grill, Eva Solo Outclass
Claus Jensen and Henrik Holbaek,
Tools Design
Porcelain, aluminium,
stainless steel and bamboo
H: 19cm (7¹⁄₂in)
Diam: 30.5cm (12in)
Eva Denmark, Denmark
www.evasolo.com

Zaha Hadid, Futuristic kitchen products

The Z-Island kitchen is the latest in Zaha Hadid's investigations into the concept of fluidity, a theme she first developed in the Ice-Series, which took its inspiration from the observation of natural phenomena such as the melting of ice or the flow of glaciers and their moraines. Known for her dramatic architectural designs, Hadid has designed a kitchen that is a futuristic and visionary re-interpretation of an old design staple, the kitchen island. Created in collaboration with DuPont Corian and Ernestomdia kitchen specialists, the 'intelligent' environment responds to our senses of vision and touch while at the same time meeting a variety of functional requirements. A multimedia centre includes sound actuators and LEDs which enable the user to surf the internet, listen to music or create personalized ambiences from a centralized touch

control panel. The kitchen consists of two free-standing units, one associated with 'fire' and the preparation of food, the other with 'water', which houses the sink and dishwasher. A wall-mounted modular cabinet system, with morphologically shaped doors set within a series of squares that have been rotated in different directions, provides space for storage and appliances and creates a series of complex patterns. 'The overall environment has been designed to be responsive to the user's needs – it can be adjusted to address the sense in many ways, translating the intellectual into the sensual while emphasizing the complex and continuous nature of the design. The user is given the unique opportunity to experiment with unexpected and totally immersive environments,' explains Hadid.

Kitchen, Z. Island
Zaha Hadid
Corian Glacier White
Main island: H: 180cm (70⅞in),
W: 80cm (31½in), L: 450cm (177⅛in)
Second island: H: 90cm (35⅜in),
W: 160cm (63in), L: 120cm (47¼in)
Wall panels: 100 pieces, each
W: 60cm (23⅝in), L: 60 (23⅝in)
Company Hasenkopf. Germany
www.hasenkopf.de

Ballpoint pen, Vivo
Konstantin Grcic
Stainless steel
L: 13.1cm (5⅛ in)
Diam: 1.1cm (½ in)
Lamy, Germany
www.lamy.com

Fountain pen and ballpoint pen,
Studio
Hannes Wettstein
Palladium finish or matt-black
lacquer finish
L: 13.8cm (5½in)
Diam: 1.2cm (½in)
Lamy, Germany
www.lamy.de

Mobile phone, Infobar
Naoto Fukasawa
Magnesium alloy, polycarbonate,
ABS, acrylic
H: 13.8cm (5½in)
W: 4.2cm (1⅝in)
D: 1.1cm (⅜in)
KDDI Corporation,
Casio Hitachi Mobile
Communications Co., Japan
www.au.kddi.com

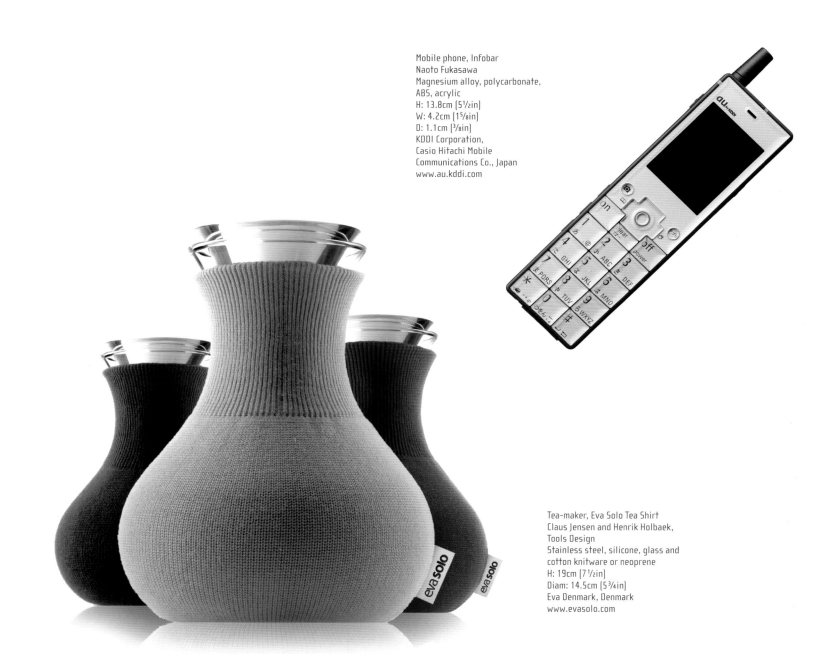

Tea-maker, Eva Solo Tea Shirt
Claus Jensen and Henrik Holbaek,
Tools Design
Stainless steel, silicone, glass and
cotton knitware or neoprene
H: 19cm (7½in)
Diam: 14.5cm (5¾in)
Eva Denmark, Denmark
www.evasolo.com

Kitchen, Place
Foster & Partners
Various materials: stainless steel,
aluminium, lacquered glass,
lacquered crystal, (brown) oak
Various dimensions
Dada, Italy
www.dadaweb.it

Radiator, Common Fence
Lucy Merchant
Powder-coated stainless steel
H: 48cm (19in)
W: 95cm (37¼in)
D: 6cm (2⅜in)
Prototype
Lucy Merchant, UK
www.lucymerchant.com

Radiator, Scudi
Massimo Iosa Ghini
Steel
H: 72.5cm (28½in)
W: 169cm (67in)
D: 13cm (5in)
Antrax IT, Italy
www.antrax.it

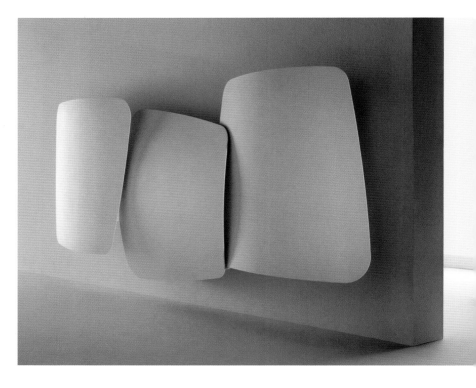

Radiator, Square
Ludovica and Roberto Palomba
Aluminium, steel
Vertical large: H: 200cm (79in),
W: 61cm (24in)
Vertical small: H: 140cm (55in),
W: 31cm (12¼in)
Square: H: 56cm (22in),
W: 56cm (22in)
Tubes Radiatori, Italy
www.tubesradiatori.com

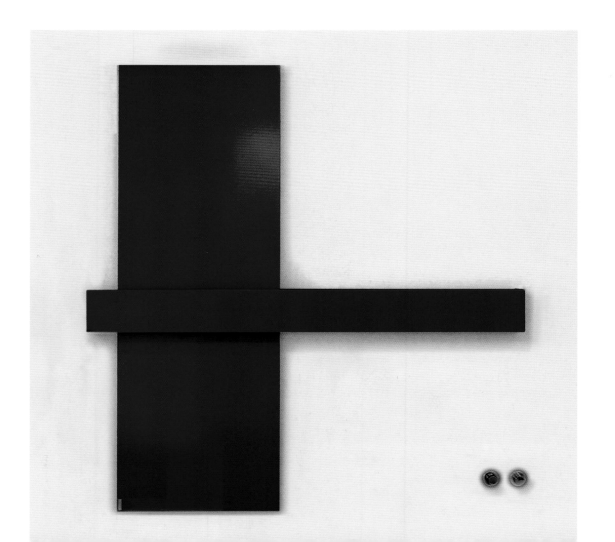

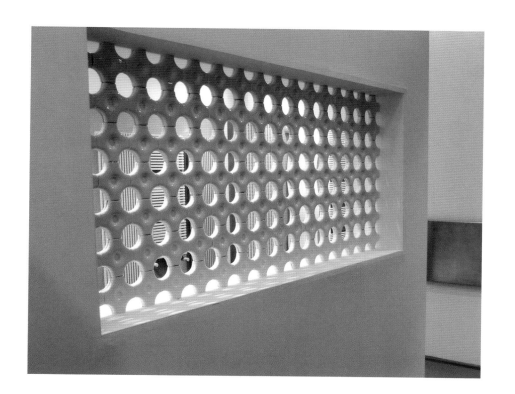

Satyendra Pakhalé, Add-On modular radiator

Satyendra Pakhalé's modular radiator was conceived in ceramic but is now mass-produced in aluminium to comply with safety regulations. 'I wanted to develop a radiator that had personality and would fit into an interior with its own modest dignity... ceramic seemed to be a great choice as it maintains its temperature and radiates heat for a long time. Add-On is a fresh, innovative way of looking at a typology that has been boring and neglected. I am glad at the way it has turned out, especially because I could take my first idea of making a radiator with its own distinction and personality, controlling all the stages of design and product development/product engineering to achieve the result I wanted. This is the kind of project I really love to work on, as it involves issues that are dear to my heart, such as technological and manufacturing challenges, and creating fresh, innovative, utilitarian products.'

Radiator, Add-On
Satyendra Pakhalé
Die-cast aluminium
H: 102cm (40in)
W: 175cm (69in)
D: 8cm (3⅛in)
Tubes Radiatori, Italy
www.tubesradiatori.com

Sewing machine, Flat Mode
Itay Potash
Stainless steel, aluminium,
plastic, polyurethane rubber
H: 4cm (1⅝in)
W: 28cm (11in)
L: 38cm (15in)
Prototype
Israel

Shoe, Origarment
Marie O'Connor
Leather, screen-printed
silk, rubber
W: 28cm (11in)
D: 2.5cm (1in)
Limited edition for
oki-ni, London
Evisu, UK
www.evisu.com

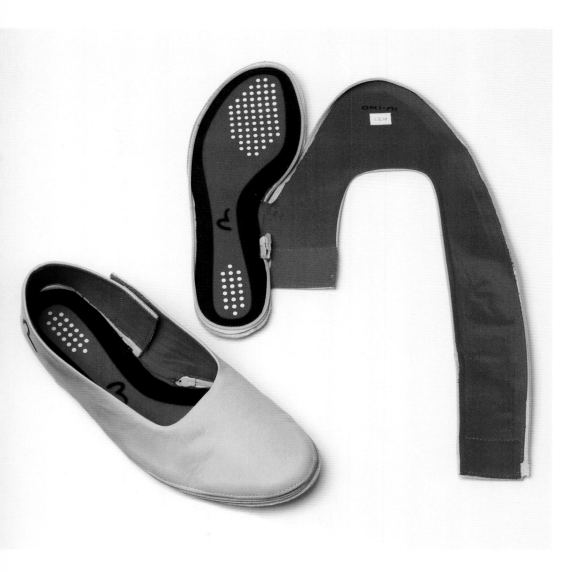

Itay Potash, Flat Mode sewing machine

Bezalel Academy of Arts and Design graduate Itay
Potash re-examines a long-overlooked typology.
Standard sewing machines are usually heavy work-
horses that have to be dragged out of the cupboard
when needed; the Flat Mode is a sleek, attractive
and easily portable reinterpretation no larger than
a laptop computer.

Potash recognizes that today, unless you are a
seamstress, the family sewing machine is needed
only every now and then for minor repairs, or
the odd pair of curtains. Influenced by hours of
watching his grandfathers using cumbersome and
complicated machines in their Tel Aviv tailor shops,
he wanted to create a user-friendly version. Flat
Mode has only four buttons, one to lift the arm up
and down, one to select the desired programme,
a third to select a stitch pattern and the last to
re-set the machine. When the task is completed,
the arm folds in and around the body, the pedal
is detached and the whole thing is stored away in
a protective bag.

Marie O'Connor, Origarment shoe

With the beauty of Oriental Origami, the two-
dimensional form is zipped together to form a
shoe. The designer, Marie O'Connor, added a secret
Braille message on the insole, which serves as
grip for footing.

Scott Henderson, Polder Z-Series ironing board
Breathing new life into a boring and banal object,
Scott Henderson has redesigned the ironing board.
Apart from using a heat-resistant silicone pad and
adding a rail around the generous cradle to hang
shirts, the real innovation is in the overall shape,
replacing the Y-shaped tubular leg structure with a
Z shape (so-called because its stability comes from
the way the legs crisscross on a diagonal). Adding
to this the fact that electrical wires run up through
the board into a power outlet under the top surface,
it means that tripping over ironing leads and legs
is a thing of the past.

Ironing board, Polder Z-Series
Scott Henderson
Stamped sheet metal, injection-moulded
polypropylene, silicon
H: 92cm (36¼in)
W: 140cm (55⅛in)
D: 39cm (15⅜in)
Polder, USA
www.polder.com

Shoe, Zvezdochka
Marc Newson
Plastic
Nike, USA
www.nike.com

Cufflinks, Gresham Blake
Marc Boase
Injection-moulded rubber
H: 1cm (⅜in)
W: 6cm (2⅜in)
L: 10cm (4in)
Naiad Plastics, UK
www.naiadplastics.com

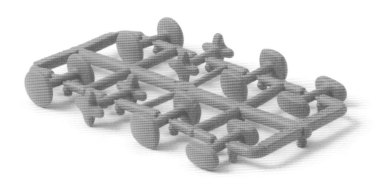

3PART design team, Cheetah wheelchair

The Cheetah wheelchair is specially designed for children. Simon Skafdrup developed an interest in design for special-needs groups while studying industrial design at the Aarhus School of Architecture, Denmark. The Danish industrial design company R82 visited the school to take a look at recent student projects and were so impressed by Skafdrup's work that they commissioned him to design the Cheetah.

The product took two and a half years to develop and the process began by interviewing children, both disabled and not, for their views, likes and dislikes. The information gathered has resulted in a wheelchair that has more of the boy racer about it than any connotation with disability. The name comes from the power and speed of the animal after which it is named and is expressed in the chair's form: the three wheels make it not only easier to handle but permits it to take up less space and is 'cool' for kids. Fully adjustable both horizontally and vertically, the chair adapts as the child grows. The tubular wheel frame is made from aluminium while the rest is of fibre-enhanced plastic, which is lightweight yet strong. A detachable tray is made from Plexiglas. Accessories are often difficult to design attractively, but this addition is both durable and discreet.

Children's wheelchair, Cheetah
3PART design team
Fibre-enhanced plastic
Frame: aluminium tubes, stainless-steel screws and PA6 parts
Cushions: fire-resistant foam
Cover: fire-resistant comfort fabric
H: 76–80cm (29⅞–31½in)
W: 58–70cm (22⅞–27½in)
L: 90cm (35⅜in)
R82, Denmark
www.R82.com

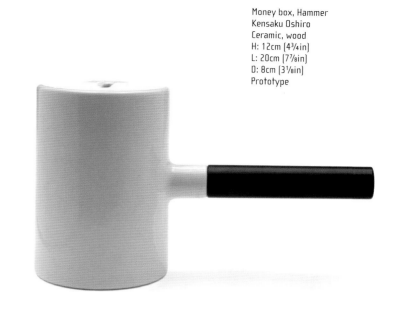

Money box, Hammer
Kensaku Oshiro
Ceramic, wood
H: 12cm (4¾in)
L: 20cm (7⅞in)
D: 8cm (3⅛in)
Prototype

Watch, Vakio watch SILAK 002
Harri Koskinen
Strap: leather
Case: stainless steel
Band: urethane
L: 40.6cm (16in)
W: 3cm (1⅛in)
Issey Miyake / Seiko Instruments,
Japan
www.seiko.co.jp

Pushchair, Bugaboo
Max Barenbrug
High-performance plastics,
synthetic rubber,
aluminium, nylon
H: 97–113cm (38⅛–44½in)
W: 91.5–104cm (36–41in)
D: 60cm (23⅝in)
Limited edition of 1000
Bugaboo International,
The Netherlands
www.bugaboo.com

Teague and Samsung, Digital Projector

In collaboration with Samsung, Teague has developed a portable digital projector concept, which received the 2006 iF Product Design Award. The projector is no larger than a digital camera and is intended for the highly mobile professional who needs ease-of-use, portability and connectivity at his or her fingertips. The system offers the possibility of interacting and controlling projection via a mobile phone by way of the latest laser diode technology.

Projector, Portable Digital Projector
David Wykes, Stefan Pannenbecker and Anthony Smith
of Teague; Jeff Higashi of Samsung Electronics
Aluminium, injection-moulded plastic and glass
H: 17.7cm (7in)
W: 3cm (1⅛in)
L: 17.7cm. (7in)
Prototype
Samsung, USA
www.samsung.com

Tom Stables, Hulger projector

The Hulger Projector is the graduation project of Tom Stables, who left Central Saint Martins College of Art and Design in 2006. The handy device projects an enlarged image of a digital camera's display screen onto any flat surface. Stables was last year's runner-up at London's Business Design Centre's New Designer of the Year competition; the judges cited the Hulger as being a 'very resolved piece... (which) demystifies technology, making it accessible for use by anyone'.

Projector, Hulger
Tom Stables
H: 6.5cm (2⅝in)
W: 6.5cm (2⅝in)
D: 7.8 or 11.8cm (3 or 4⅝in)
Prototype
UK

Projector, Com Projector
Toshihiko Sakai
ABS, aluminium
H: 12.5cm (5in)
W: 5.6cm (2¼in)
D: 5.8cm (2⅝in)
Comfortable Communications Co.,
Japan
www.com2.jp

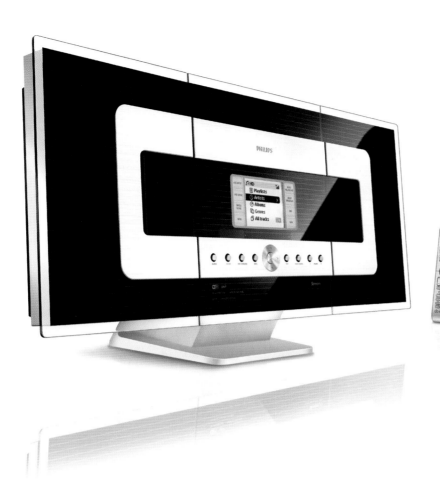

Philips Wireless Music Centre

Just as 2005 was the year of the MP3 player, 2006 seems to be flooded with major electronic companies bringing out their versions of the wireless music centre. We have illustrated the Sonos offering on page 121 and feature here Philips' design. With a slightly sci-fi, techno look, it is much less discreet than its competitor, but what it might lose in aesthetics it gains in technology. Equipped with a 40GB hard drive, it's able to wirelessly stream music around your house to up to five satellite units. Up to 750 CDs of music can be loaded on it using the built-in CD player or via the Ethernet port, which is also used to pull down information on the ripped CDs from the Gracenote CDDB database.

Audio system, Wireless Music
Centre and Station WACS700
Philips Design Team
Moulded ABS, painted
H: centre: 30.5cm (12in);
station: 29cm (11³/₈in)
W: centre: 60.8cm (24in);
station: 36cm (14¹/₈in)
D: centre: 17.5cm (7in);
station: 12.8cm (5in)
Royal Philips Electronics,
The Netherlands
www.philips.com

Television, TAV-L1
Shoichiro Matsuoka, Phillip Rose,
Nick Foster, Naofumi Yoneda
and Masami Nitta (Sony)
H: 47cm (18¹/₂in)
W: 85.6cm (33⁵/₈in)
L: 106.9cm (42in)
SONY Corporation, Japan
www.sony.net

Antonio Citterio, Kinesis exercise system

Antonio Citterio has earned a reputation for his uncompromising design; his self-coined term for this is 'proto-rationalism'. Citterio's approach, whether designing a piece of furniture or the interior of a boutique hotel, starts by examining the thought process behind the function. Citterio graduated in architecture from the Polytechnic of Milan in 1976 and was greatly inspired by the Modern movement with its adage of 'Less is More'. His work today is still concerned with rationality, functionalism and the philosophy of living with comfort and practicality. 'The nature of my approach – reflected by the work of my practice – lies in the intimate relationship between product design, interior design and architecture.' His assistant jokes that, with an individual floor devoted each to architecture, interiors and products, even the design layout of his own studio reflects this plurality.

Despite his prolific output, designing for design's sake is something that Citterio cannot comprehend. Aesthetics are not enough; there has to be a technological development or a good reason for an object, and clarity of tectonics or social need that makes a product or an interior usable. Citterio's latest project is a bit of a departure but, given his philosophy, it is not totally surprising that he was attracted to the commission.

Technogym is an Italian company that manufactures high-end fitness equipment and is known not only for its technological expertise but also for the importance it places on aesthetics. Using premium materials and concealing all machinery, its products are adaptable to the home environment. Technogym approached Citterio to design the housing for its new Kinesis exercise system, which comprises grips, cables and weight stacks that are wall-mounted and provide resistance-based training. Citterio has set the cable supports in a free-standing panel, which is covered with a hard-wearing, leather-like material. 'Citterio understands how to combine design and functionality in a piece which will not look out of place in anyone's home,' comments Tim O'Connell, the marketing director of Technogym.

Home exercise system,
Kinesis Personal
Antonio Citterio
Oakwood, inox, polished
stainless steel 1810
H: 210cm (83in)
W: 170cm (67in)
D: 115cm (45³/₈in)
Technogym, Italy
www.technogym.com

Leica Rangemaster CRF 1200

The Leica CRF 1200 has a range of up to 1,100 metres (3,609ft). At a touch of a button the rangefinder will tell you precisely how far away your objective is and the sensitive scan mode ensures precise recordings no matter how difficult the target might be. An LED display adjusts itself to surrounding light conditions.

Laser Rangefinder, Leica Rangemaster CRF 1200
Leica Camera
Carbon-fibre-reinforced plastic, painted finish
H: 7.5cm (3in)
W: 11.3cm (4½in)
D: 3.4cm (1¼in)
Leica Camera, Germany
www.leica-camera.com

Olympus SP-350 digital camera

The SP-350 doesn't look good. Let's face it, it's downright ugly. It's big, solid, chunky and heavy. The electronics expert Cliff Smith rather amusingly wrote in his review, 'If this camera were a car it would be a 1989 Toyota Hi-Lux pickup with a bull bar, off-road tyres and a cement mixer in the back. If you painted it matt green and stencilled on a few serial numbers it would look like a piece of military hardware.'

Yet of its kind, a semi-professional digital camera, it is way ahead of its field, with many cutting-edge features and an image quality that is second-to-none. Its main selling feature is its wide-angled lens, which allows for greater photographic versatility. It also has a choice of seven scene programme modes, including two underwater settings for use with the optional waterproof case, illustrated here – the only redeeming aesthetic feature.

Digital camera, Olympus SP-350
Jun Tkahashi, Olympus Imaging Design Team
High-density plastic
H: 6.5cm (2½in)
W: 9.95cm (3⅝in)
D: 3.5cm (1⅜in)
Olympus, Japan
www.olympus.com

Wall plate/switch,
Pom Pom Dimmers
Margaret Orth, PhD
Stainless steel and
polyester yarns, machine
embroidery, designtex fabric,
light dimmer
H: 12cm (4¾in)
W: 7.6cm (3in)
D: 3.8cm (1½in)
Limited-batch production
International Fashion
Machines, USA
www.ifmachines.com

Wallbox, Lyneo Preset
Joel Spira
Plastic and metal
H: 8.6cm (3⅜in)
W: 8.6cm (3⅜in)
D: 3.6cm (1⅜in)
Lutron Electronics Co., USA
www.lutron.com

Wall-plate switches, Timeless
Bart Spillemaeckers
Various metals (stainless steel,
brass, bronze, etc.)
Single: H: 8.4cm (3⅜in),
W: 8.4cm (3⅜in), D: 0.5cm (¼in)
Double: H: 17.5 (6⅞in),
W: 8.4cm (3⅜in), D: 0.5cm (¼in)
Lithos, Belgium
www.lithos-sb.be

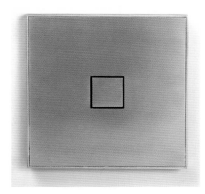
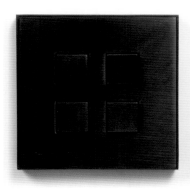
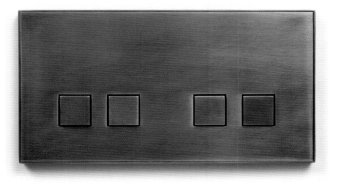

Electrical sockets, Unplugged
Barbara de Vries
Plastic
Various dimensions
Prototype
www.barbaradevries.com

Wall-mounted humidifier and heater, Petra
Marta Laudani and Marco Romanelli in
collaboration with Marcello Pinzero
Fine porcelained stoneware
H: 12.5 or 22.5cm (4⅞ or 8⅞in)
L: 31.5cm (12⅜in)
D: 4cm (1⅝in)
Limited-batch production
Laboratorio Pesaro, Italy
www.laboratoriopesaro.com

Mobile phone, Media Skin
Tokujin Yoshioka
Prototype
KDDI, Japan
www.kddi.com

Mobile phone, Pebble
Motorola Consumer
Experience Design Team
Glass, plastic, polycard sheet
stock, stainless-steel metal
injection moulding,
die-cast zinc, chrome
W: 4.9cm (2in)
L: 8.65cm (3⅜in)
D: 2cm (¾in)
Motorola, USA
www.motorola.com

Nokia 7280 mobile phone

At first glance you could be forgiven for not
knowing what the Nokia 7280 is, let alone how it
functions. The mobile phone's pen-shaped art deco
design is by far one of most outrageous Nokia has
yet conceived. The display is used horizontally with
the screen (which doubles up as a mirror when not
in use) to the left and the Navi Spinner to the right.
There is no keypad, and the only buttons are the
upper and lower soft keys, the 'call' and 'end' keys
and the middle selection pad. The spinner, which can
be moved clockwise or counter-clockwise, is the
only way to dial a number, type an SMS or navigate
through the menus. The earpiece is located above
the screen and is lined in suede.

Mobile phone, Nokia 7280
Nokia Design Team with Tej Chauhan,
Grace Boicel and Tanja Fisher
Injection-moulded polyurethane ABS,
metal and suede detailing
H: 11.5cm (4½in)
W: 3cm (1⅛in)
D: 1.5cm (⅝in)
Nokia, Finland
www.nokia.com

Radio, Radius
Sam Hecht & Ippei Matsumoto,
Industrial Facility
Hi-pressure ABS injection
moulding, magnetic headphones
H: 13cm (5in)
W: 1.8cm (¾in)
L: 1.8cm (¾in)
Lexon, France
www.lexon-design.com

Radio, Silver Radio
Konstantin Grcic
Sterling silver
H: 2cm (¾in)
W: 30cm (11⅞in)
L: 45cm (17¾in)
De Vecchi, Italy
www.devecchi.com

AM/FM travel radio, Match Radio
Gabriele Pezzini
Extruded aluminium, injected ABS
H: 7.5cm (3in)
W: 2.5cm (1in)
L: 10cm (4in)
Areaplus, China
www.areaplus.com

Loudspeakers
Front Design
Glass
H: 45cm (17¾in)
D: 30cm (11¾in)
Prototype
Sweden
www.frontdesign.se

Phone headset, Jawbone
Yves Béhar
Injection-moulded ABS, aluminium
L: 5cm (2in)
Aliph, USA
www.aliph.com

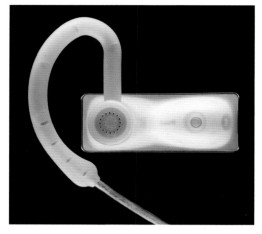

Headphones, QUALIA010
Satoshi Suzuki
Magnesium-alloy, carbon filter
Custom made
Sony Corporation, Japan
www.sony.net

Sonos Digital Music System

The Sonos Digital Music System is a stylish way to stream music from your computer. A portable, wireless jukebox, it is no bigger than an average radio, and its streamlined aesthetic means it can sit unobtrusively in any environment. All you have to do is supply the speakers, connect one of the players to your digital music library (on a MAC, PC or NAS box) and up to thirty-two hubs can be controlled by the wireless controller. Each 'zoneplayer', as Sonos refers to them, has a range of 45.7m (150 ft) and delivers a 50-watt channel amp.

Consumer electronics, Sonos
Digital Music System
Mieko Kusano and Rob Lambourne
(Sonos)
UV-proof high-grade PC (plastics)
and aluminium
Sonos Controller 100 (CR100):
H: 16.5cm (6½in); W: 2.4cm (1in);
L: 9.7cm (3¾in)
Sonos ZonePlayer 100 (ZP100):
H: 25.9cm (10¼in); W: 11.2cm
(4½in); L: 20.8cm (8¼in)
Sonos ZonePlayer 80 (ZP80):
H: 13.7cm (5⅜in); W: 7.4cm (3in);
L: 14cm (5½in)
Sonos, USA
www.sonos.com

CD radio
Naoto Fukasawa
ABS plastic and punching metal
H: 9.5cm (3¾in)
W: 43cm (16⅞in)
D: 18cm (7⅛in)
Prototype
Plusminuszero, Japan
www.plusminuszero.co.jp

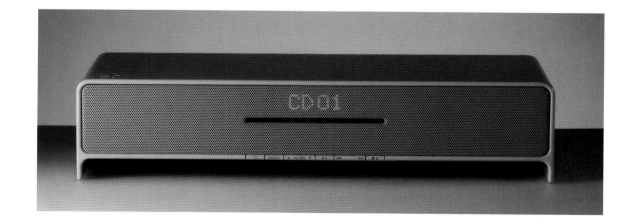

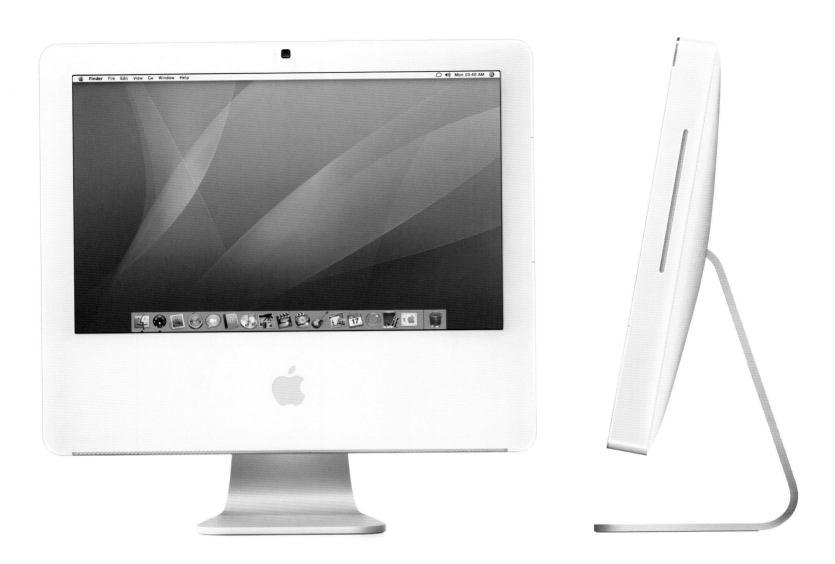

Computer, iMac
Apple Design Team
Polycarbonate/ABS
Dimensions of 17-inch model:
H: 43cm (17in)
W: 42.6cm (16¾in)
D: 17.3cm (6¾in)
Apple, USA
www.apple.com

Laptop computer, MacBook Pro
Apple Design Team
Aluminium
Dimensions of 15.4-inch model:
H: 2.6cm (1in)
W: 35.7cm (14in)
D: 24.3cm (9⅝in)
Apple, USA
www.apple.com

Computer mouse, Mighty Mouse
Apple Design Team
Polycarbonate/HDPE
H: 3cm (1⅛in)
W: 6cm (2⅜in)
L: 11cm (4⅜in)
Apple, USA
www.apple.com

Power Mac G5 FG
(finished goods) box
Apple Design Team
Double wall B/C flute
H: 71cm (28in)
W: 32.4cm (12⅞in)
L: 59cm (23¼in)
Apple, USA
www.apple.com

iPod Shuffle
Apple Design Team
Polycarbonate/ABS
H: 0.8cm (¼in)
W: 2.5cm (1in)
L: 8.4cm (3¼in)
Apple, USA
www.apple.com

Apple iPod Shuffle Sport Case
Apple Design Team
Polycarbonate/ABS
H: 1.5cm (⅝in)
W: 4cm (1⅝in)
L: 10.5cm (4in)
Apple, USA
www.apple.com

Apple iPod Shuffle Dock
Apple Design Team
Polycarbonate/ABS
H: 2.89cm (1in)
W: 7.9cm (3in)
D: 5.8cm (2¾in)
Apple, USA
www.apple.com

Apple iPod Nano Armband
Apple Design Team
Polyurethane, spandex,
stainless steel
W: 9.5cm (3¾in)
L: 42cm (16½in)
Apple, USA
www.apple.com

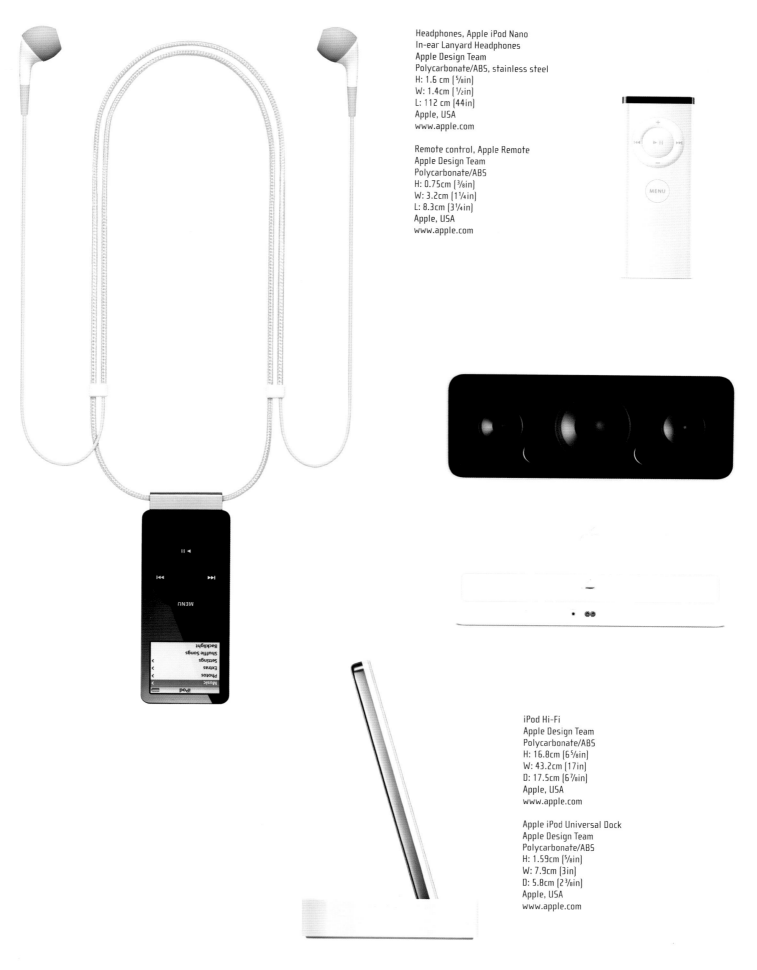

Headphones, Apple iPod Nano
In-ear Lanyard Headphones
Apple Design Team
Polycarbonate/ABS, stainless steel
H: 1.6 cm (⅝in)
W: 1.4cm (½in)
L: 112 cm (44in)
Apple, USA
www.apple.com

Remote control, Apple Remote
Apple Design Team
Polycarbonate/ABS
H: 0.75cm (⅜in)
W: 3.2cm (1¼in)
L: 8.3cm (3¼in)
Apple, USA
www.apple.com

iPod Hi-Fi
Apple Design Team
Polycarbonate/ABS
H: 16.8cm (6⅝in)
W: 43.2cm (17in)
D: 17.5cm (6⅞in)
Apple, USA
www.apple.com

Apple iPod Universal Dock
Apple Design Team
Polycarbonate/ABS
H: 1.59cm (⅝in)
W: 7.9cm (3in)
D: 5.8cm (2⅜in)
Apple, USA
www.apple.com

Michael Young

2006 was Michael Young's year. Since his graduation from Kingston University in London in 1992, he has built a formidable reputation as a designer of great vision, whose use of bold shapes and colours, mixed with interesting combinations of materials, has produced works that are at once aesthetically simple yet technically involved. 2005, however, saw the culmination of a string of projects: a bike for Taiwan-based multi-national company Giant, a range of accessories for Schweppes, designs for Magis and, in Taipei, the first of a series of Corian-clad interiors for the Dr James clinic (he is currently working on a second in Florence and has the commission for a chain worldwide). This output will guarantee that this modest man, who of his own admission 'lives in design exile', will not be allowed to remain insular.

Moreover, his foresight is not restricted to being able to gauge what the market demands aesthetically. While European designers and manufacturers are currently obsessed by what they see as the threat of China, Young has embraced Asia's rise as a major design and production powerhouse by uprooting his family and relocating, at first to Taiwan and then Hong Kong, to work directly with manufacturers in mainland China. Recognizing that this emerging industry lacks a basic knowledge of design, he knew that he had something to offer: a way of working with designers from Europe to get rid of the association of the Chinese being 'copiers' and build up an independent reputation without which they will never be able to export successfully.

Young says of himself, 'When I moved to Taipei, the industrial-design fraternity took a liking to me. Because I came from Europe they saw me as an unusual sort of character. They still call me a fashion designer but they understand that this is what's needed. I have two companies: Michael Young, which produces all my furniture and interiors, etc., and I am also the creative director on an industrial design company. I come up with the ideas of what I consider to be a European style, or whatever, and they supply the technical expertise. I have a great admiration for Starck, who opened up the vision between industry and thinkers, and I still strongly believe that the role of the designer today is to create purposeful objects that generate economy and inspire others to do so.'

Of a huge concern to Young is the effect manufacturing is having on global warming. Young considers China to be getting a bad press, given that America is responsible for twenty-five per cent of the pollution of the planet. There are deep-rooted and historical reasons for this, but China's present ecological

standards are far more stringent than those of the USA. However, the problem is international. He says: 'I'm doing a lot of work on it at present to find out how I can be a good and caring designer and I would like to impress how utterly important it is for other designers and design manufacturers to join me in this. My company makes designs on products that can be recycled wherever possible; we specify plastics that have earth-friendly pigments and stabilizers and make sure that we are fully aware of what other additives are in the materials we use.'

Michael Young's Gel Mouse is a new and ergonomic interpretation of the traditional computer mouse pad. Brightly coloured, with a strong emotional aesthetic, the curved plastic lower shell holds a soft and 'friendly' PU pad with touch-sensitive controls. The pad is much more considerate to the joints of the hand, the upper padded segment providing an easier surface than the hard forms of the traditional mouse.

Computer mouse, Gel Mouse
Michael Young with Ignition Asia
Ergo gel and ABS
H: 2.5cm (1in)
W: 5.8cm (2¼in)
L: 10cm (4in)
OEM
www.ignition.com

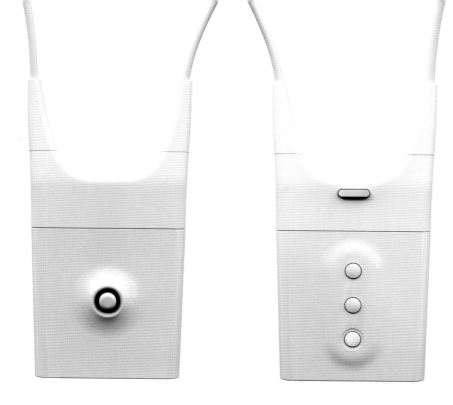

MP3 player with integrated
headphones, Mica
Norway Says (Torbjørn Anderssen,
Andreas Engesvik, Espen Voll)
Plastic casing
H: 5.4cm (2⅛in)
W: 4cm (1⅝in)
D: 1.6cm (⅝in)
Asono, Norway
www.asono.com

Optical mouse, Mouse Trap
Sam Hecht, Industrial Facility
ABS injection moulding
H: 3cm (1⅛in)
W: 5.2cm (2in)
L: 8.3cm (3¼in)
Lexon, France
www.lexon-design.com

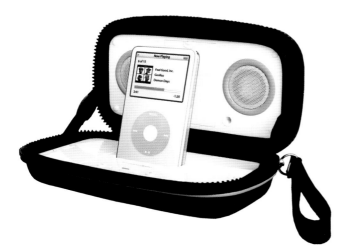

Suitcase for iPod, iFusion
Team Sonic Impact
Polycarbonate/ABS interior,
ballistic nylon exterior,
compression-moulded EVA case
H: 6.5cm (8½in)
W: 16.5cm (6½in)
L: 21.5cm (8½in)
Sonic Impact Technologies, China
www.si5.com

Wireless electronic display,
Toughbook CF-08
Panasonic
Magnesium-alloy casing
Screen: 26.4cm (10⅜in)
Panasonic (Matsushita Electric
Industrial Co.), Japan
www.toughbook-europe.com

Printer, Wall-mounted Printer
Gwendolyn Floyd
Plastic, printer components
H: 34cm (13³⁄₈in)
W: 31cm (12¹⁄₄in)
D: 5cm (2in)
Prototype
Ransmeier and Floyd,
The Netherlands

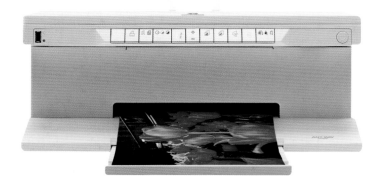

Printer, Any Way
James Irvine and Alberto Meda
H: 19.6cm (7³⁄₄in)
W: 46.2cm (18³⁄₈in)
D: 33.9cm (13³⁄₈in)
Olivetti, Italy
www.olivetti.com

Photo printer, My Way
IDEO
H: 22cm (8⁵⁄₈in)
W: 16.2cm (6³⁄₈in)
D: 9cm (3¹⁄₂in)
Olivetti, Italy
www.olivetti.com

Pop-up card, Bloom cards
Tahmineh Javanbakht and
Clara von Zweigbergk
Paper and plastic
H: 14cm (5½in)
W: 14cm (5½in)
D: 2.5cm (1in)
Artecnica, USA
www.artecnicainc.com

Chicken coop, Eglu
Omlet (James Tuthill,
Johannes Paul, Simon Nicholls
and William Windham)
Polyethylene
H: 80cm (31½in)
W: 115cm (45¼in)
L: 300cm (118in)
Omlet, UK
www.omlet.co.uk

Eglu chicken coop

As more and more thirty-something, upwardly mobile middle classes are leaving their urban areas for a more pastoral existence, those that cannot try and add a little of the country to their city environments. Eglu is constructed in polyethylene and can accommodate two medium-sized hens. It consists of a slatted roosting area, nesting box, sliding droppings tray and a back-door egg port.

Shower, Rain Sky M
Michael Sieger
Steel
H: 230cm (91in)
W: 161.2cm (63½in)
L: 146.2cm (57½in)
Aloys F. Dornbracht & Co., Germany
www.dornbracht.com

Tap, Pan Plus
Ludovica and Roberto Palomba
Chrome, brass finish
Zucchetti, Italy
www.zucchettidesign.it

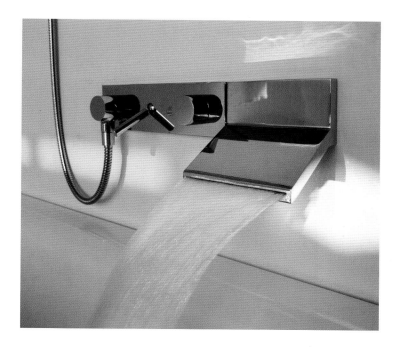

Washbasin, Consolle 01
Ludovica and Roberto Palomba
Ceramic
W: 50cm (19⅝in)
L: 160cm (63in)
Laufen, Switzerland
www.laufen.com

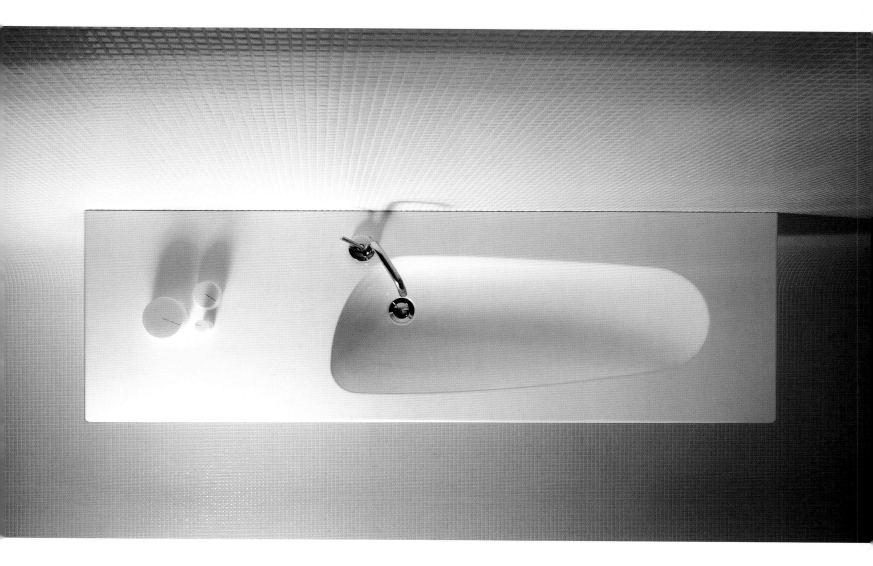

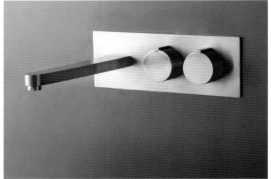

Bath and kitchen taps, W1
Norbert Wangen
Brushed stainless steel AISI 304
Precise dimensions undefined
Prototype
Boffi, Italy
www.boffi.com

Washbasin, Zone
Piero Lissoni
Corian
L: 120 or 240cm (47 or 94½in)
W: 47cm (18½in)
H: 15 or 30cm (5⅞ or 11¾in)
Boffi, Italy
www.boffi.com

Bathtub, Terra
Naoto Fukasawa
Cristal Plant
H: 55cm (21⅝in)
W: 171cm (67in)
D: 151cm (59in)
Boffi, Italy
www.boffi.com

Tap, MEM4
Michael Sieger
Brass
H: 27.6cm (10⅞in)
W: 25.5cm (10in)
D: 16.5cm (6½in)
Limited-batch production
Aloys F. Dornbracht & Co.,
Germany
www.dornbracht.com

Washbasin, Miniwash 25
Guilio Cappellini
Ceramic
H: 32cm (12⅝in)
W: 25cm (9⅞in)
D: 40cm (15¾in)
Ceramica Flaminia, Italy
www.ceramicaflaminia.it

Two-hole mixer with
extensible spray and
cover plate, Elio-2 hole
Sieger Design
Coated brass, polished
chrome or platinum
matt finish
H: 34.2cm (13½ in)
Diam: 23.5cm (9¼in)
Aloys F. Dornbracht & Co.,
Germany
www.dornbracht.com

Bathtub, Exline
Benedini Associati
Exmar (composite material made
of resin and quartz powder)
H: 55cm (21⅝in)
L: 170cm (67in)
Agape, Italy
www.agapedesign.it

Washbasin, Roto
Benedini Associati and
Maurizio Negri
Polypropylene
H: 75cm (29½in)
W: 16cm (6¼in)
Agape, Italy
www.agapedesign.it

Showerhead and mixer tap,
Istanbul Collection
Ross Lovegrove
Chrome-plated brass
Limited-batch production
VitrA, UK
www.vitra.co.uk

Mixer tap, Love Me
Maurizio Duranti
Chrome-plated brass
H (inc. lever): 17.8 or 36cm
(7 or 14¼in)
W: 6.5cm (2½in)
IB Rubinetterie, Italy
www.ibrubinetterie.it

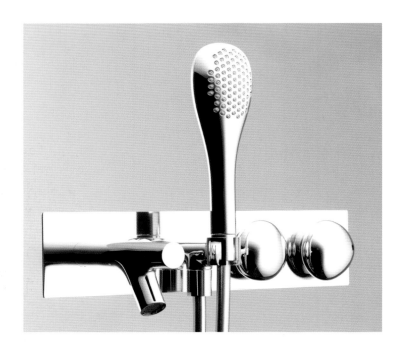

Free-standing single-lever
bath-shower mixer tap, Pan Soft
Ludovica and Roberto Palomba
Chrome-plated brass and
components resistant to corrosion
and scaling
H: 77.3cm (30½in)
L: 20cm (7⅞in)
W: 5cm (2in)
Diam: 5cm (2in)
Mass production
Zucchetti Rubinetteria, Italy
www.zucchettidesign.it

Shower tray, Geo Tray Tondo
Ludovica and Roberto Palomba
Acrylic
Diam: 80cm (31½in)
Kos, Italy
www.kositalia.com

Washbasin and stand,
Bathroom Programme
Jaime Hayón
Table under washbasin: lime-wood
legs and water-repellent MDF top,
gloss polyurethane lacquer finish
Washbasin and accessories:
stoneware porcelain, colour and
precious metal (gold and platinum)
enamel finish
Basin stand: H: 81cm (31⅞in);
L: 140cm (55⅛in)
Basin: W: 47cm (18½in);
L: 65.5cm (25¾in)
Artquitect, Spain
www.artquitect.net

Jaime Hayón

The designs of Jaime Hayón are an acquired taste but share a freshness and originality which make them hard to ignore. They are infused with a Spanish love of colour, flamboyance and the surreal, tempered – if that's not a contradiction in terms – with a playfulness, as well as a technical know-how, which elevates them from art pieces to commercial design. Today Hayón is one of the fastest-rising industrial designers working in his native country of Spain, although, reluctant to be associated with what he considers to be still a very small and emerging design scene, there are plans afoot for a relocation to London.

Hayón was born in Spain to working-class parents, and his love of art and design arose from the skateboarders he hung out with and for whom he created the graphics for their decks and boards. Hayón's gutsy style was noticed by a major American manufacturer which sponsored him to travel to San Diego, California, where he became involved with the underworld of graphic design and graffiti art connected to the subculture of skateboarding, an experience he says still informs his work today.

He returned to Madrid to study industrial design at the newly formed Istituto Europeo di Design, where he received a very formal education. Upon graduation he was employed by the think-tank institute, Fabrica, Benetton's Communcation and Applied Arts Foundation in Italy, where within a year he was promoted to head of their design department, conceiving concepts for exhibitions, store interiors and products. In 2004 he moved to Barcelona and began to get noticed with a series of plastic toys, the Qee collection, which received a cult following, especially in Japan. Installations and exhibitions of porcelain objects followed – non-functional works, which allowed him to explore his inventiveness while at the same time giving him experience in working with high-quality materials and different manufacturing processes.

The 'Mediterranean Digital Baroque' show and 'Mon Cirque' installation drew the attention of bathroom manufacturer Artquitect, which commissioned the collection of bathroom furniture, the sink console of which is illustrated here. This colourful series of mirrors and lights, which are integrated into sink pedestals and baths but which can be bought separately, move away from a minimalist aesthetic but are not as overtly baroque as Marcel Wanders' Soap Star series for Bisazza.

The analogy to Wanders and the appropriation of historical motifs is an annoying one for Hayón, who maintains that his inspirations are varied and that the importance of the bathroom set lies elsewhere. 'If you see how many new things this bath gave to the market – it gave colour, it gave luxury materials like gold and platinium, the fact that you could buy any piece individually, the integration of light within the toilet...'

Showtime is Hayón's latest collection and was launched during the Milan Furniture Fair 2006 to much acclaim. The range includes a hooded outdoor plastic chair and accompanying ornamental sofa, ceramic robotic vases (page 181) and a multilegged lacquered wooden sideboard, which take their inspiration from the theatricality and fun of the Hollywood musical. 'I'm not just looking for stories, but I want my furniture to make an impact, surprise people and make them use their imagination.'

Washbasin mixer tap,
Axor Starck X
Philippe Starck
H: 28cm (11in)
Plateau-style tap head:
W: 15cm (6in); L: 15cm (6in)
Hansgrohe, Germany
www.hansgrohe.com

Sink/washbasin, Pot Sink
Inci Mutlu and Luca Milano
Glazed reinforced terracotta
H: 45cm (17⅝in)
W: 38cm (15in)
D: 43cm (17in)
Diam: 38cm (15in)
Droog Design,
The Netherlands
www.droogdesign.com

Bath, Soap Stars
Marcel Wanders
Cristal plant
H: 58.6cm (23in)
W: 200cm (79in)
D: 144cm (57in)
Bisazza, Italy
www.bisazza.com

Bathtub, Aqhayon Bath
Jaime Hayón
Wood, fibreglass,
ceramic accessories
H: 50cm (19⅝in)
W: 100cm (39¾in)
L: 205cm (81in)
Artquitect, Spain
www.artquitect.net

Sink, H2O
Lorenzo Damiani
Plastic
H: 33cm (13in)
W: 34cm (13⅜in)
D: 48cm (19in)
Limited-batch production
Lorenzo Damiani Studio, Italy
lorenzo.damiani@tin.it

Bath Tub, BA10 AOMORI
Philipp Mainzer and Johana Egenolf
European oak, stainless steel
H: 90cm (35⅜in)
W: 135cm (53⅛in)
L: 135cm (53⅛in)
e15, Germany,
in collaboration with
Duravit, Germany
www.e15.com
www.duravit.de

Washbasin/sink, Starck X Washbasin
with Metal Console
Philippe Starck
Ceramic, metal, high-gloss colour
fired into the ceramic
H: 80–85cm (31½–33½in)
W: 57cm (22½in)
L: 110cm (43¼in)
Duravit, USA
www.duravit.com

Ceramic washbasins,
Voyage to the Orient
Paola Navone
Ceramic, hand-formed
and hand-painted
Diam: 46cm (18⅛in)
Rapsel, Italy
www.rapsel.it

Counter-top washbasins with
decoration, Impronte patterns
on Flow (FL11, FL13)
Terri Pecora
Ceramic
FL11 (square): L: 39.5cm (15½in),
W: 39.5cm (15½in), H: 13cm (5⅛in)
FL13 (oval): L: 51.5cm (20¼in),
W: 31.5cm (12⅜in), H: 13cm (5⅛in)
Simas, Italy
www.simas.it

139

lighting

Lamp, Princess
Christophe Pillet
Polyethylene
H: 100cm (39³/₈in)
Diam: 75cm (29¹/₂in)
P. Serralunga, Italy
www.serralunga.it

Lighting, Atomium
Hopf and Wortmann, Büro für Form
PE, steel
6 x max. 40W/E14 bulb
H: 60cm (23⁵/₈in)
W: 60cm (23⁵/₈in)
L: 60cm (23⁵/₈in)
Kundalini, Italy
www.kundalinidesign.com

Stick lighting, Blown ups
Thelermont Hupton
Glass lamp with steel stand
G9 halogen bulb
H: 140cm, 55¹/₈in
Limited-batch production
Thelermont Hupton, UK
www.thelermonthupton.com

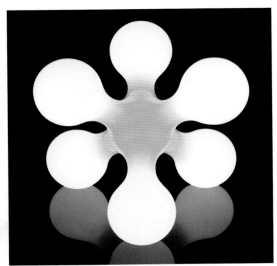

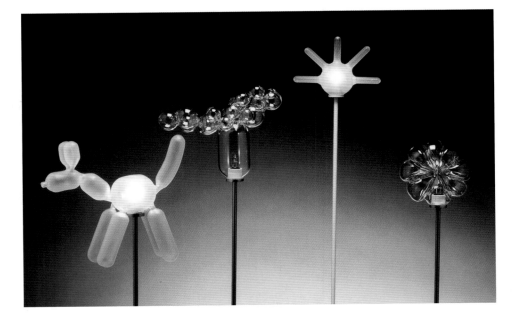

Ross Lovegrove, System X lighting

Since 1923 Yamagiwa has led the field as one of Japan's major manufacturers of high-profile lifestyle products from furniture to electronics and home appliances, but it is probably in the field of lighting that it is best known internationally, having the worldwide distribution rights for the Frank Lloyd Wright collection. Euroluce 2005 saw the company rebranding its long-established firm with a range of specially designed lights by five of the most technologically innovative designers working today:

Ross Lovegrove, Naoto Fukasawa, Tokujin Yoshioka, Shigeru Uchida and Shiro Kuramata. Lovegrove's System X ceiling light explores his fascination with architectonic design. A re-invention of the fluorescent strip light, it is structural, flexible and modular, offering the possibility of being configured to produce geometric systems from one to an infinite number of units. The pieces can be connected in two directions thanks to rubber joints, creating a big grill, a line or different sized circles.

Lighting, System X
Ross Lovegrove
Plastic, aluminium,
2 x T421W fluorescent lamps
1 x 60W silver bowl lamp
H: 20.5cm (8in)
Diam: 33cm (13in)
Yamagiwa, Japan
www.yamagiwa.co.uk

Lighting, System X
Ross Lovegrove

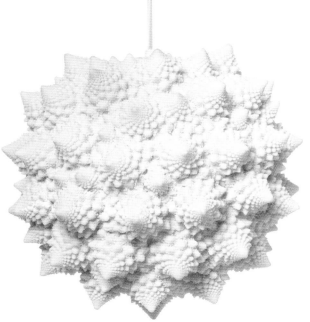

Ulrika Jarl, Romanesco pendant light
Structures and patterns from nature, the rhythm and repetition of related shapes are an important influence in Jarl's work as a designer/maker. But though her forms, like those of nature, often appear simple, they contain highly complex and detailed elements. Her lighting is decorative rather than purely functional, altering the light not merely reflecting it. The intricate Romanesco light is made up of more than 100 polyester resin casts, each one of a different Romanesco cauliflower.

Pendant light, Romanesco
Ulrika Jarl
Polyester resin and fibreglass
LED
H: 35cm (13¾in)
Diam: 35cm (13¾in)
Habitat (UK), UK
www.habitat.net

Pendant light, L'Eclat Joyeux
Ingo Maurer
Porcelain, metal
6 x 60W; 1 x 250W halogen bulb
H: 120cm (47¼in)
One-off
Ingo Maurer, Germany
www.ingo-maurer.com

Lamp, Liquid Light PX02
Masahiro Fukuyama
Glass, E27G-95 Hal, max 100W
H: 20cm (7⅞in)
W: 30cm (11¾in)
Metalarte, Spain
www.metalarte.com

Floor lamp, Nanoo5
Christian Deuber
Steel, polyethylene
3 x 60W/E27 bulb
H: 180cm (71in)
Diam: 40cm (15¾in)
Nanoo by Faserplast,
Switzerland
www.nanoo.ch

Floor lamp, Alma
Lela Scherrer, Christian Deuber,
Jörg Boner
Haute couture textiles, wood
1 x 100W halogen bulb
H: 165cm (65in)
W: 55cm (21⅝in)
D: 35cm (13¾in)
One-off

Table lamp, HANA
Reiko Okamoto and Marcia Iwatate
Base: cast iron, urethane rubber or
hand-shaped glass
Shade: 'washi' mulberry paper,
mouth-blown glass, cotton muslin
Suspension: 2mm piano wire
1 x 20W incandescent clear lamp
Various dimensions
Prototype
Marei, Japan
www.marei-ltd.com

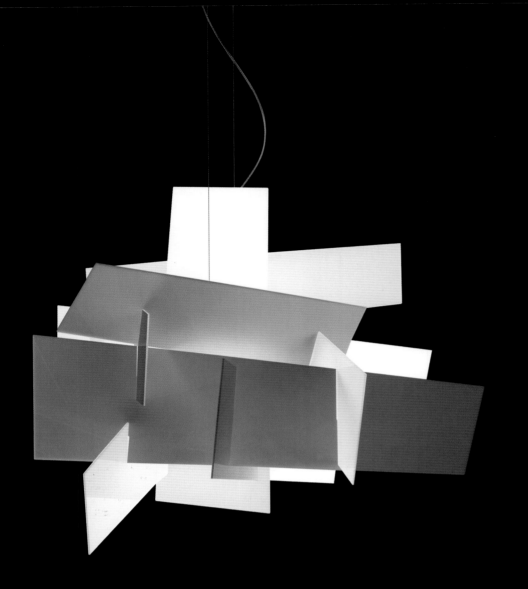

Suspension lamp, Big Bang
Enrico Franzolini and
Vicente Garcia Jiménez
Methacrylate
1 x 200W halogen bulb
H: Max 200cm (79in)
L: 96cm (38in)
Foscarini, Italy
www.foscarini.com

Lighting fixture, Collage Pendant
Louise Campbell
Shade: Laser-cut acrylic
Suspension: natural anodized
aluminium
Max. 100W/E27 matt
incandescent lamp
H: 36cm (14⅛in)
Diam: 60cm (23⅝in)
Louis Poulsen Lighting, Denmark
www.louis-poulsen.co.uk

Floor light, Fat Spot
Tom Dixon
Prismatic polycarbonate, copper
H: 35cm (13¾in)
Diam: 30cm (11⅞in)
Tom Dixon, UK
www.tomdixon.net

Chandelier, Dear Ingo
Ron Gilad
Powder-coated steel
16 x 25W/E27 bulb
H: 80cm (31½in)
W: 100cm (39⅜in)
Moooi, The Netherlands
www.moooi.com

Philippe Starck, Guns lamp collection

People are very much divided on whether they like what Philippe Starck is doing or not. Nowhere is that more evident than the reaction provoked by his collection of gold-plated Kalachnikov floor, table and bedside lamps for Flos. Whether you love them or hate them, one thing is definite: you will not be indifferent (although Martí Quixé was quoted in *Domus* as saying 'It is just another lamp').

Has Starck gone too far in mixing opulence and murder? Design isn't art, the guns aren't one-off sculptures, they are not decommissioned weapons transformed. These lights are objects that are going to be bought in their hundreds by only the wealthy, and neither Flos not Starck will have any power over what point will be made by their acquisition. Should design get involved with politics – should it have that function? If these lamps weren't designed by Starck, would we feel the same way about them? If it were some unknown, would we say he/she had definitely transgressed the narrow borders of good taste? Hani Rashid, writing for *Domus*, states 'When design privileges innovation, culture speeds along; when art drives forward relevant social commentary, culture flourishes. When one mixes the two, it is effectively the combination of salt and gasoline. The motor splutters to a deliberate and irreversible death.'

Range of table and floor lamps,
Guns collection
Philippe Starck
Die-cast zamak, chrome-plated
ABS stem
1 x 75W/E27 bulb (Bedside)
1 x 150W/E27 bulb (Table)
1 x 250W/E27 bulb (Lounge)
Bedside: H: 30cm (11¾in)
Table: H: 75cm (29½in)
Lounge: H: 170cm (67in)
Flos, Italy
www.flos.com

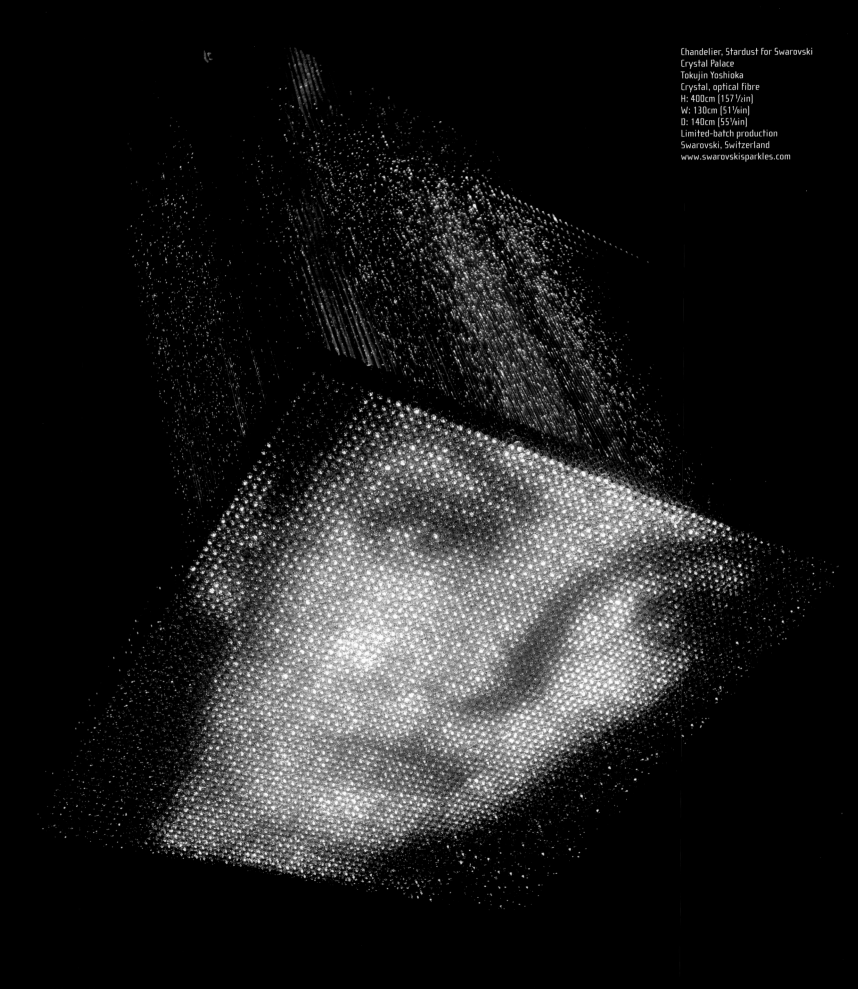

Chandelier, Stardust for Swarovski
Crystal Palace
Tokujin Yoshioka
Crystal, optical fibre
H: 400cm (157½in)
W: 130cm (51⅛in)
D: 140cm (55⅛in)
Limited-batch production
Swarovski, Switzerland
www.swarovskisparkles.com

Swarovski, Crystal Palace chandeliers

There is something fascinating in all that glitters. Swarovski has long been associated with kitsch, albeit expensive, crystal ornaments, from playful kittens to swans gliding across frozen pools of reflective glass, which, trapped in showcases in jewellery and gift shops around the world, catch the artfully angled spotlights to refract a myriad of rainbow colours. It's the kind of spectacle your eye is drawn to but you pretend to ignore.

So how come well-known designers are now striving to be associated with the brand, vying to be selected as the next batch creating works of magic and beauty to be displayed at the launch of the annual Crystal Palace collection, an exhibition of chandeliers enchanting visitors at each year's Milan Furniture Fair? The answer lies in the persuasive hand of Nadja Swarovski, the great-great-granddaughter of the company's founder, Daniel Swarovski, the

Bohemian crystal maker who moved to Wattens in Austria in 1895 to set up his eponymous business. Returning from New York where she worked as a fashion PR, her aim was to liberate the creative potential of a material she fell in love with as a child and which she felt, in these post-minimal days, was ready for a contemporary reassessment.

Since 2002 Nadja has been the brain behind the transformation of her family's brand by inviting names such as Ron Arad, Tom Dixon, Toord Boontje, Ingo Maurer and the Campana brothers to come up with chandeliers that push the boundaries not only of aesthetics but also of technical innovation. Last year's collection saw interpretations by Ross Lovegrove, Naoto Fukasawa, Hussein Chalayan, Gaetano Pesce, Michael Gabbellini, Jurgen Bey, Ineke Hans, Basso & Brooke, Simon Heijdens, Laurene and Constantin Boym and Chris Levine. Returning to the original concept of the range, the

lights marry advanced technology with contemporary design, using nature as an inspiration.

Taking influence from the science-fiction world, Tokujin Yoshioka's Stardust projects moving 3D images, accompanied by sounds, onto hundreds of round, 20mm-diameter (³/₄in) crystals suspended on individual fibres. Each radiates light independently and sways with full freedom of movement.

At first sight, Pesce's chandelier, Mediterraneo, looks traditional but it's not long before you notice that it's moving slowly and seamlessly, replicating the graceful dance of a jellyfish through water, exuding ever-changing colour and radiance as it does so.

Chandelier, Mediterraneo for
Swarovski Crystal Palace
Gaetano Pesce
140 crystal strands, each
consisting of 87 Swarovski
crystals illuminated by LEDs
H: 304cm (120in)
Diam: 203cm (80in)
Swarovski AG, Switzerland
www.swarovskisparkles.com

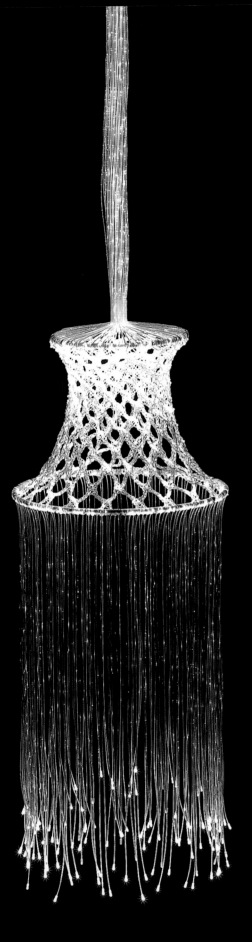

Lamp, Bobbin Lace Lamp
Niels van Eijk
Glass fibre optic
100W halogen bulb
Diam: 80cm (31½in)
Quasar Holland, The Netherlands
www.quasar.nl

Light Fixture, Ether
Patrick Jouin
Structure: polished steel
Diffusers: blown glass
4 x 75W G53 bulb (+ 1 x 5W
multicoloured high-power LED)
H: 86 or 143cm (33⅞ or 56¼ in)
Diam: 100cm (39⅜in)
Murano Due, a brand division of
Firme di Vetro, Italy
www.muranodue.com

Patrick Jouin

Patrick Jouin was born in 1967 in the small French town of Mauves-sur-Loire. Of his own admission he comes from a simple background (his father is a wood and metal turner and his mother a nurse) and had no idea about design until the age of seven, when he won a drawing competition run by the local bank. The prize was to be flown to Paris to spend a day in the Beaubourg Museum. Jouin describes the experience as an epiphany and from that moment on he decided he wanted to be designer.

Jouin's early life was influential. From his father he inherited a love of technology and a respect for materials. From his mother he learned the importance of studying the body like a machine, which he believes is essential in his career as an architect and designer as it allows him to perceive space in relation to human needs and demands. Jouin studied industrial design at ENSCI (Les Ateliers), graduating in 1992, and was immediately employed by the Compagnie des Wagon Lit, working on the interiors of restaurant cars. A period at Tim Thom, Thomson multimedia, under the artistic direction of Philippe Starck, was followed by a placement in Starck's studio, where he remained for four years while at the same time producing his own lines for VIA, receiving their Carte Blanche invitation (a grant given to promote young designers) in 1998. He opened his own studio the following year.

Jouin says that having worked for Starck is a double-edged sword. In the light of his own considerable achievements – he was recently described as the designer of the most beautiful restaurants in the world – it is an annoyance that he is still sometimes referred to as one of Starck's *enfants*, but he nonetheless recognizes that he gained a lot from the experience. Not only did he develop connections with leading furniture manufacturers – Cassina, Kartell and Alessi – early on in his career, but, more importantly, he learned the lesson that an idea is never finished, that a designer should always be struggling to make something better, even at the final stages.

For Jouin, designing is an intellectual game with the idea being as important as the finished product. Jouin works both as product designer and interior architect, and in both disciplines experiments with materials and technology to push the boundaries of what is possible. Dotted around this book are examples of his recent products; the most important of which are his pioneering researches into the process of stereolithography, a technique that enables the translation of digital models directly into 3D objects.

The Solid series and the One Shot stool for Materialise will be considered, retrogressively, as events in the history of contemporary design. In his interiors he considers himself to be neither a designer nor an architect but a DJ of space, orchestrating poetry, mood and ambience to create environments that are luxurious yet functional, traditional yet modern and contemporary yet elegant.

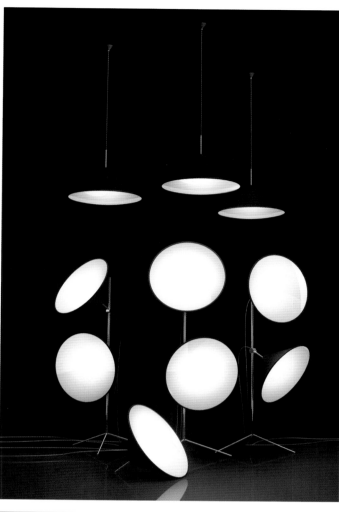

Multifunctional lamp,
Cone Light
Tom Dixon
Aluminium, acrylic
1 x incandescent 100W bulb
H: 42cm (16½in)
W: 22 or 74cm (8⅝ or 29⅛in)
Tom Dixon, UK
www.tomdixon.net

Pendant lamp, Caravaggio
Cecilie Manz
Drawn steel,
high-gloss enamel
1 x E27
H: 32.5cm (12¾in)
Diam: 25.7cm (10⅛in)
Lightyears, Denmark
www.lightyears.dk

Lamp, Cleo
Flavio Mazzone and
Claudio Larcher,
Modoloco Design Milano
Metal tubing with
transportation
grasp handle
1 x 40W
H: 25cm (9⅞in)
W: 50cm (19⅝in)
L: 25cm (9⅞in)
Prototype
spHaus, Italy
www.sphaus.it

Chandelier, Lech
Isabel Hamm
Glass rods, polished stainless steel,
halogen bulb
1 x 50W halogen bulb
H: 72cm (28⅜in)
Diam: 55cm (21⅝in)
Limited-batch production
Isabel Hamm Gestaltung, Germany
www.isabel-hamm.de

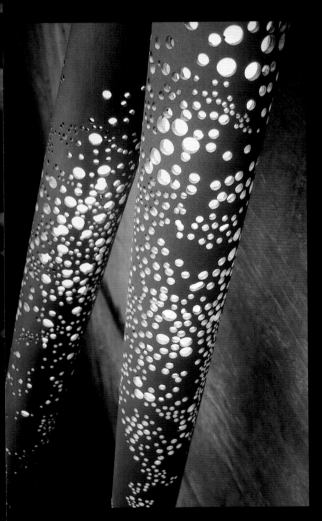

Pieke Bergmans

Pieke Bergmans studied graphics at the Academie of Arts, St. Joost, Breda, and 3D design at HKA Art-School, Arnhem, The Netherlands, before graduating from the Design Academy in Eindhoven. The academy has produced many of the designers we associate with Droog. Founded as an art school, since the 1980s it has taught purely design but is well known for the conceptual-led, autonomous designers it produces. However, often criticized for being uncommercial, the school is now endeavouring to leaven this idea-led approach to design with a strong dose of industrial design knowledge in how things are made. Interviewed for *Icon* magazine's January 2006 edition, Miriam van der Lubbe, an Eindhoven tutor, was quoted as saying 'Previously there was such a big gap between the school and the real world that it wasn't working well. The skills had gone and in their place came concept. But you need more than just concept'.

Bergmans' work is a fine balance between the poetry we have come to expect from the contemporary Dutch school and a sound understanding of what it means to exist realistically in the competitive world of design. Although her work is playful, she combines function with form and message, working closely with different industries and factories, and in different materials to explore the possibilities of various production facilities. 'The goal is to make personalized mass production where irregularities are ruled in. The next step is to control imperfection to create objects which are more interesting and individual but which can be conceived commercially.'

Her Virus series, which includes organic fungus-shaped mirrors that crawl along the walls and floors, and a stick light that looks as if it has been eaten away by a metal-consuming insect, seek to bind the many disciplines and materials in which she works. Her products are called viruses due to their natural forms and the way that they come to life, spread and take over, adapting to various conditions, disrupting common ideas and the predictable evolution of design. If mass production is analogous to the reproduction of the healthy cell, then Pieke's designs allow for change and serendipity, creating processes in which no one object is quite the same as the last – a mutation of the cell into a contagious disease, which can be creative as well as destructive.

Light, Metalworm
Pieke Bergmans
Aluminium tube, stick light
H: 150cm (59in)
Diam: 5cm (2in)
Prototype
www.piekebergmans.com

Chandelier, Come Rain Come Shine
Tord Boontje
Steel rods wrapped with ribbon;
crochet piece; organza, cotton
and silk
H: 111.8cm (44in)
Diam: 55.9cm (22in)
Limited-batch production
Artecnica, USA
www.artecnicainc.com

Lighting, Clear
Ayala Serfaty
Clear skin webbed over glass
structure, with ceramic base
H: 70cm (27½in)
W: 80cm (31½in)
D: 20cm (7⅞in)
One-off
Aqua Creations, Israel
www.aquagallery.com

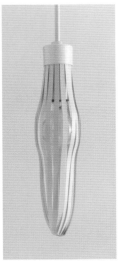  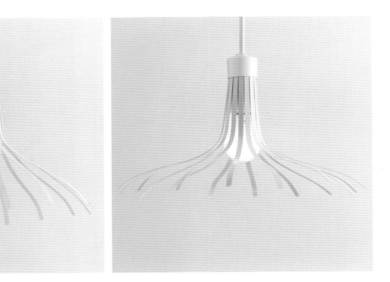

Lighting, Hanabi (above)
Oki Sato Nendo
Shape-memory alloy,
incandescent lamp
H: 78cm (31in)
W: 26cm (10¼in)
D: 26cm (10¼in)
Prototype

Lighting, HBM (below)
Carlotta de Bevilacqua
Modules: in aluminium and in PETG
with aluminized and translucent film
Union elements: in PETG, structure
and bottom plate in chrome metal
Max. 2 x 150W/E27 bulbs
Various dimensions
Danese Milano, Italy
www.danesemilano.com

Floor lamp, Dandelion
Richard Hutten
Laser-cut powder-coated steel
1 x 100W/E27 bulb
H: 170cm (67in)
W: 80cm (31½in)
Limited-batch production
Moooi, The Netherlands
www.moooi.com

Pendant lamp, Attalo
Wilmotte & Associés
Aluminium, methacrylate,
3 x 80W fluorescent RGB
light source (G 5) – T16
H: 6cm (2³⁄₈in)
L: 21cm (8¹⁄₄in)
W: 152cm (60in)
Artemide, Italy
www.artemide.com

Pendant lamp, Light Frame (right)
Stephen Burks
Acrylic
H: 17.4 or 30cm (6⅞ or 11¾in)
Diam: 72 or 100cm (28⅜ or 39⅜in)
David Design, Sweden
www.david.se

Pendant lamp, Zeppelin (below)
Marcel Wanders
Cocoon, powder-coated steel,
injection-moulded PMMA
1 x 60W/E27 bulb (Size 1)
1 x 150W bulb (Size 2)
Diam: 55 or 110cm (21⅝ or 43in)
Flos, Italy
www.flos.com

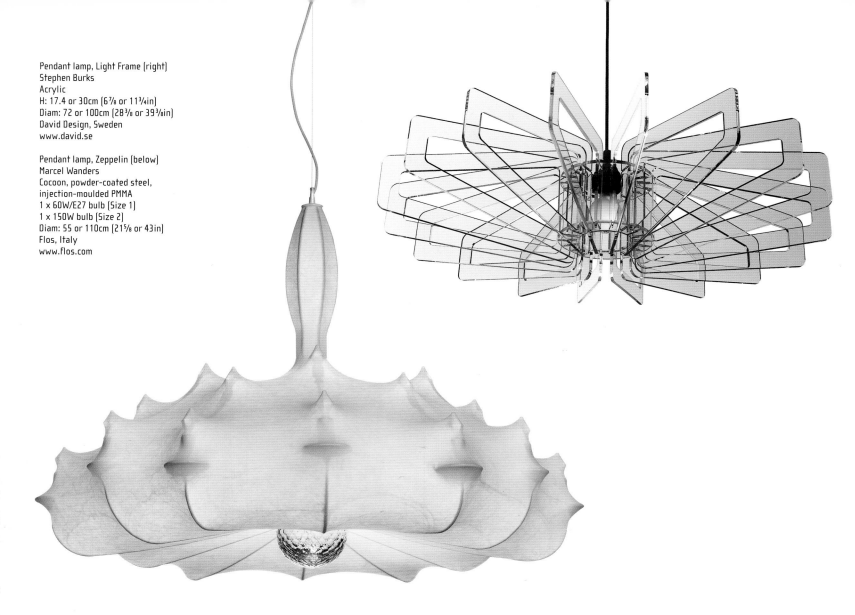

Marcel Wanders, Zeppelin pendant lamp

In 1960 Achille Castiglioni and Pier Giacomo
Castiglioni produced their Cocoon Lamps made of
metal and sprayed fibreglass for Flos. In 2005
Marcel Wanders paid tribute to this groundbreaking
design with Zeppelin, a pendant lamp made of steel
and pressed crystal cover. Each lamp starts as
white-painted steel, sprayed twice with a composite
of resin and spun polymer fibre, which sticks to the
framework to become a taut membrane. Once set,
a coat of protective transparent paint is added, the
entire process taking ninety-six hours.

Lamps, Agave
Diego Rossi and
Raffaele Tedesco
Injection-moulded,
transparent methacrylate
21, 26, 32 or 42W
fluorescent bulb
Diam: 17, 26 or 70cm
(6¾, 10¼ or 27½in)
Luceplan, Italy
www.luceplan.com

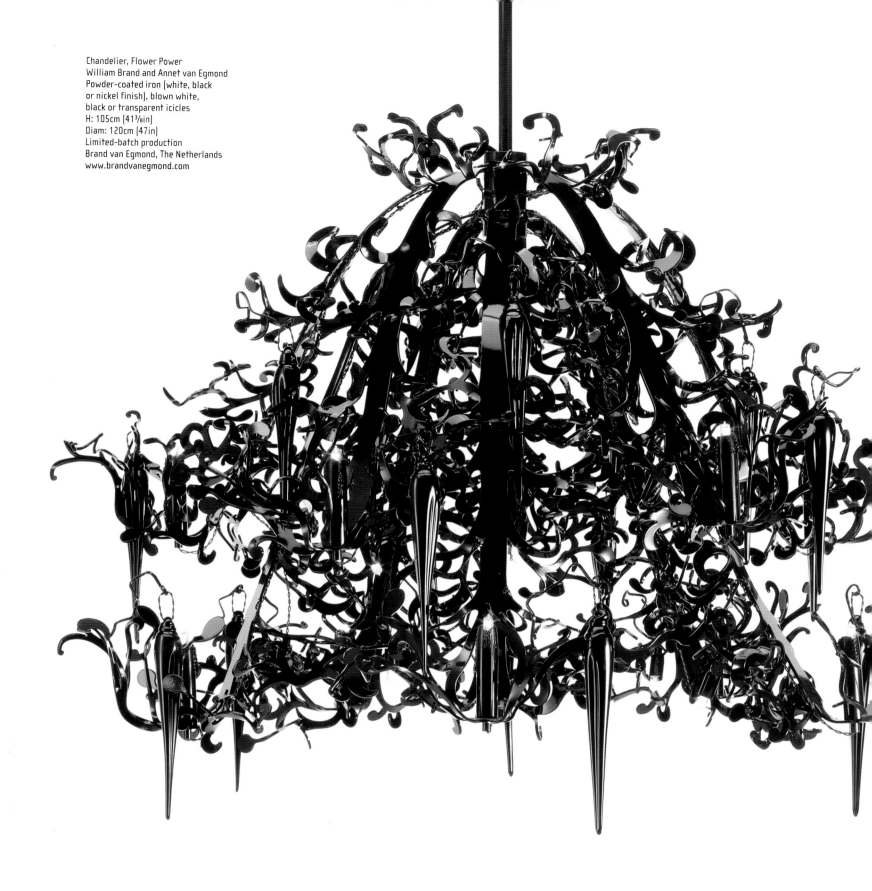

Chandelier, Flower Power
William Brand and Annet van Egmond
Powder-coated iron (white, black
or nickel finish), blown white,
black or transparent icicles
H: 105cm (41³/₈in)
Diam: 120cm (47in)
Limited-batch production
Brand van Egmond, The Netherlands
www.brandvanegmond.com

Chandelier, Vortexx
Zaha Hadid and Patrick Schumacher
Moulded fibreglass, thermo-shaped
acrylic, car paint
High-pressure LEDs
H: 160cm (63in)
Diam: 170cm (67in)
Sawaya & Moroni, Italy
www.sawayamoroni.com

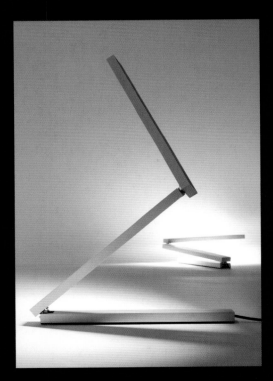

Frank Oehring, Zac light

Frank Oehring studied fine art in Essen, Germany and his work crosses the boundaries between art, applied art and industrial design. He is well known for his stage designs and kinetic light sculptures. With its many adjustable parts, the Zac light can be unfolded, stretched, extended, angled, laid down, tipped and stood upside down.

Table lamp, Zac
Frank Oehring
Metal parts with satin matt chrome
1 x T2 FM 8W bulb
H: 6–82cm (2³/₈–32¹/₄in)
W: 5cm (2in)
L: 40cm (15³/₄in)
ANTA Leuchten, Germany
www.anta.de

Richard Sapper, Halley desk lamp

The lighting company Lucesco was founded in 2004 in Silicon Valley, California, and the Halley is their first product. Using the latest LED technology, it was developed in just over a year and follows in the path of the iconic Tizio, which Richard Sapper designed in 1972 for Artemide, and is likely to become an equal milestone in the lamp design industry. Sapper was familiar with LEDs through his work in computer design for IBM but he had not heard of the new generation of new white lights, which are now being manufactured to compete with the white incandescent light of halogen bulbs. He was further convinced of the benefits of using energy-saving LEDs (they have a life span of twenty years or 50,000 hours) when he was made aware that, due to the small size of the LED head, he could further refine the balancing act he had started over thirty years ago with Tizio.

The fully articulate Artemide light was conceived to shine a precise pool of halogen light onto a piece of paper while never getting in the way of the user. Halley is even more flowing and responsive than

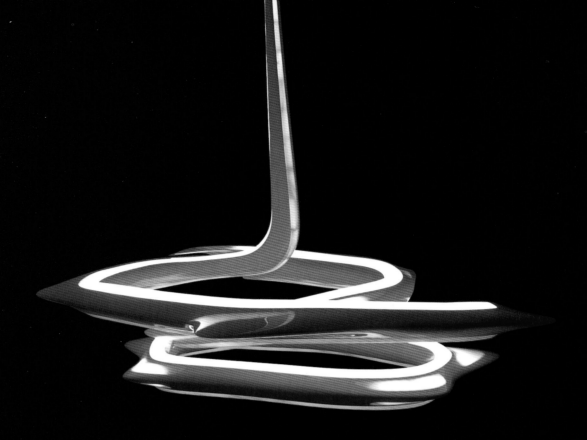

Tizio thanks to a specially patented joint, which is used at three junctures on the lamp and both connects and conducts electricity, allowing 360 degrees of motion. Friction has been added to the joints in order that the lamp does not act like a mobile, blowing freely in the breeze. The light source comprises sixteen miniature spotlights arranged in the head like soccer-stadium flood-lights. And why the name? Sapper writes, 'I thought of the name Halley because the illuminating head needs a tail for cooling purposes, because the free geometry of its movements enables it, in the orbit of its arms, to fly over the table and because the necessary cross-counterweight reminds me of a comet flying around the earth.'

Desk Lamp, Halley
Richard Sapper, Nicole Sargenti
Aluminium, steel, plastic moulding
16 x LED (3 versions)
H: 40–128cm (15³⁄₄–50³⁄₈in)
Reach: 40–120cm (15³⁄₄–47¹⁄₄in)
Lucesco Lighting, USA
www.lucesco.com

Lamps, Firefly
Frederik Roijé
Forced metal and aluminium
H: 10cm (4in)
W: 20cm (7⁷/₈in)
L: 20cm (7⁷/₈in)
Studio Frederik Roijé,
The Netherlands
www.roije.com

Desk lamp, Cloud
Jozeph Forakis
Aluminium, polycarbonate
1 x 35W halogen bulb
H: 38–68cm (15–26¾in)
Foscarini, Italy
www.foscarini.com

Candelabrum, Bold
Roderick Vos
4mm sheet metal
Little Bold: H: 40cm (15¾in)
W: 21cm (8¼in)
D: 54cm (21¼in)
Limited-batch production
Moooi, The Netherlands
www.moooi.com

Table lamp, One Line (right)
Ora-ïto
Extruded aluminium,
polycarbonate, die-cast aluminium,
1 x 11W fluorescent miniature
T2 (W4.3)
Diam (base): 16.5cm (6½in)
Artemide Design, Italy
www.artemide.it

Floor, wall or ceiling luminaire for
indoor/outdoor use, 45 ADJ FL2 (far right)
Tim Derhaag
Die-cast aluminium
1 x fluorescent bulb
H: 44cm (17³⁄₈in)
W: 8.5cm (3³⁄₈in)
L: 44cm (17³⁄₈in)
Diam: 6cm (2³⁄₈in)
Flos, Italy
www.flos.com

Tom Barker, SmartSlab textiles

Tom Barker graduated from the Royal College of Art in London, where he is now head of Industrial Design Engineering. While working with Zaha Hadid on the Mind Zone at London's Millennium Dome, they experimented with a slab of aluminium honeycomb lined with Christmas lights. Six years later Barker developed the product into a translucent, structural panel consisting of the same composite but sandwiched between clear fibreglass resin. Working in collaboration with architectural lighting distributor Targetti and LED manufacturer Cotco, he put LEDs into the hexagonal cells to create an amazing display system controlled by proprietary software, which has far-reaching potential. Users can load images, live video and graphics straight from any website. It is the world's toughest modular structural tile and can support the weight of a fully grown elephant, making it adaptable for either wall or floor use. With moving or still images, the system can be used on a huge scale (Zaha Hadid is interested in employing the technique in Tokyo Guggenheim's 600-metre (1,968ft) display – the largest video wall in the world) or domestically. The slabs, which measure 60cm (23⅝in) across, are available in two pixel sizes to allow for both interactive interior installations and oversize billboards. The slabs can be arranged in multiples of two- and three-dimensional configurations, including cubes and cylinders, and SmartSlab is presently working on a flexible 'tile', which will have the capacity to mould itself to complex curves.

Modular digital LED display, SmartSlab
Tom Barker
Polycarbonate, steel
LED
H: 60cm (23⅝in)
L: 60cm (23⅝in)
D: 15cm (6in)
Targetti Sankey, Italy
www.targetti.com

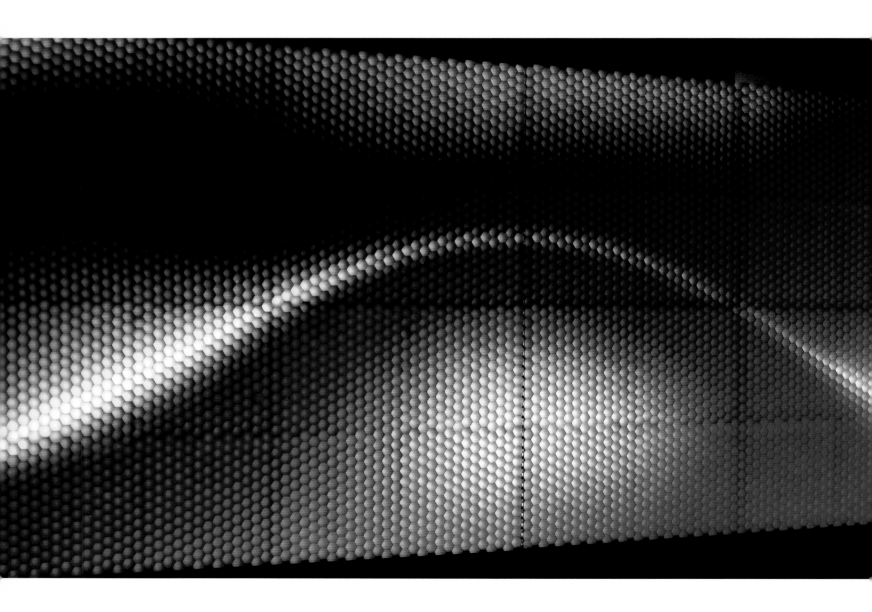

Jacopo Foggini

Jacopo Foggini explores the boundaries between art and design, producing monumental, colourful luminaires, light installations and sculptural objects. While working in the family plastics business, he discovered the possibilities of methacrylate, an industrial material used in the fabrication of car tail-lights, which has all the properties of glass but is much lighter. He invented a machine in which he heats the thermoplastic resin to 200°C (392°F), creating chromatic and versatile filaments that can be hand-moulded into phantasmagorical shapes of extraordinary beauty and luminescence.

Theatre chandelier / lighting sculpture, Il Cancelliere
Jacopo Foggini
Methacrylate
50 x halogen bulb
H: 600cm (236¼in)
Diam: 600cm (236¼in)
One-off
Jacopo Foggini, Italy
www.jacopofoggini.it

Lighting, Liana S.
Ayala Serfaty
Treated crushed silk, metal
Compact fluorescent bulbs
H: 40cm (15¾in)
W: 61cm (24in)
L: 183cm (72in)
Limited-batch production
Aqua Creations, Israel
www.aquagallery.com

Pendant lamp, Octo
Seppo Koho
Laminated birch
H: 68cm (26¾in)
Diam: 54cm (21¼in)
Limited-batch production
Secto Design, Finland
www.sectodesign.fi

Table light, Cage Light
Tom Dixon
Wire lamp in copper-finish steel
with laminated cotton shade
H: 60cm (23⅝in)
W: 30cm (11⅞in)
Tom Dixon, UK
www.tomdixon.net

Pendant fixture, Three
Mattias Ståhlbom
White, grey or black metal rod
Soft globe max. 60W/E27 bulb
H: 21.5 cm (8¼in)
Diam: 49cm (19¼in)
Zero Interior, Sweden
www.zero.se

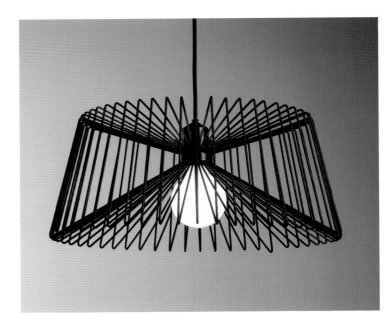

Light pendant or floor light shade,
Coral
David Trubridge
Hoop pine plywood and plastic clips
Diam: 60 or 80cm (23⅝ or 31½in)
One-off
David Trubridge, New Zealand
www.davidtrubridge.com

Lampshades, Old Fruits,
Tops and Bottoms
Bertjan Pot
Dried gourds; interior: white acrylic
paint; exterior: black acrylic paint
1 x 25W bulb
Various dimensions
Limited-batch production
Bertjan Pot, The Netherlands
www.bertjanpot.nl

Lamps, Kikomo
Renaud Thiry
PVC, fabric, steel, wood
Various dimensions
Limited-batch production
Flandesign, France

Hanging lamp, Cubrik
Antoni Arola
Nickel structure and white ABC strips
Max. 150W/E27 halogen bulb
H: 40cm (15¾in)
W: 40cm (15¾in)
L: 40cm (15¾in)
Santa & Cole, Spain
www.santacole.com

Pendant light, 24 Karat Blau
Axel Schmid
Gold, plastic, metal
1 x 60W halogen bulb
H: 21 or 29cm (8¼ or 11⅜in)
W: 33cm (13in)
L: 33cm (13in)
Ingo Maurer, Germany
www.ingo-maurer.com

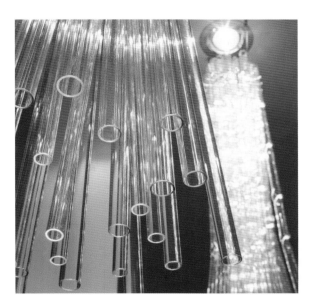

Spotlight, Fort Knox 2
Philippe Starck
Die-cast aluminium,
CDMR-111 35W bulb
L: 21cm (8¼in)
W: 18cm (7in)
D (block): 4.5cm (1¾in)
Diam: 11.5cm (4½in)
Flos, Italy
www.flos.com

Chandelier, Munich
Isabel Hamm
Glass, stainless steel
Halogen reflectors, halogen light
L: max. 150cm (59in)
Diam: max. 40cm (15¾in)
Isabel Hamm Gestaltung, Germany
www.isabel-hamm.de

Ingo Maurer, Delirium Yum table lamp

The Delirium Yum is a small version of a huge water column that Ingo Maurer created for the Kruisherenhotel in Maastricht, The Netherlands. A water spiral is generated by a propeller at the bottom of the glass vessel, which causes a sprinkling of silver dust to swirl in the vortex, mesmerizing the onlooker. 'The combination of water and light has always captivated me,' says Maurer. 'Immersed in the observation of haphazard movements, our thoughts leave their habitual course. Or is it an allegory of the restless mind, turning round and round?'

Table lamp, Delirium Yum
Ingo Maurer and Sebastian Hepting
Corian, crystal glass, carbon
fibre, silicone
1 x 35W halogen spot lamp
H: 80cm (31½in)
W: 40cm (15¾in)
L: 40cm (15¾in)
Limited-batch production
Ingo Maurer, Germany
www.ingo-maurer.com

Lighting, Jingzi (right)
Herzog and de Meuron
Silicone
1 x 150W incandescent lamp
H (head): 45cm (17¾in)
Diam: 30cm (11⅞in)
Belux, Switzerland
www.belux.com

Lighting, Cloud Series (far right)
Frank Gehry
Refined polyester
1 x 100–150W
incandescent lamp
Diam: 60cm (23⅝in)
Belux, Switzerland
www.belux.com

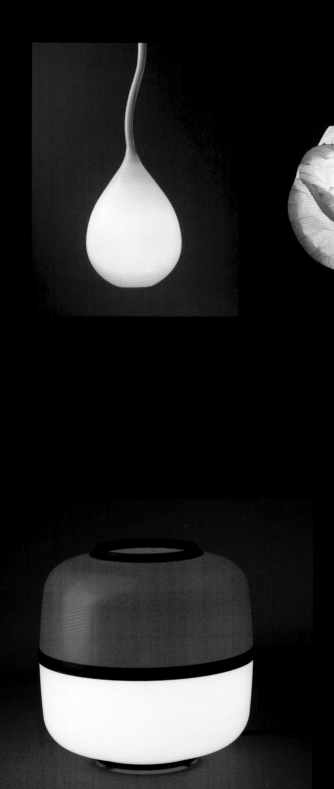

Light fixtures, Lantern
Ronan and Erwan Bouroullec
Plastic
1 x 150W/E27
Diam (diameter): 38cm (15in)
Belux, Switzerland/Germany
www.belux.com

Moisture-proof diffuser
luminaire, Scuba
Massimo Iosa Ghini
Glass-reinforced plastic,
polycarbonate, steel
H: 10cm (4in)
W: 11cm (4⅜in)
L: 130cm (51⅛in)
Zumtobel Staff, Austria
www.zumtobel.com

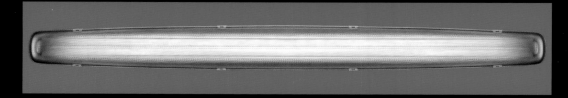

Stuart Haygarth, Tide chandelier
The Tide chandelier is made from a collection
of clear and translucent objects collected
on a specific stretch of the Kent coastline.
Each object is suspended on fishing line from
a platform above.

Chandelier, Tide chandelier
Stuart Haygarth
Clear and translucent plastic
collected from one specific
beach on the Kent coast,
fishing line, split shot,
MDF platform
1 x 100W incandescent bulb
Diam: 150cm (59in)
Stuart Haygarth, UK
stu@haygarth.abelgratis.co.uk

tableware

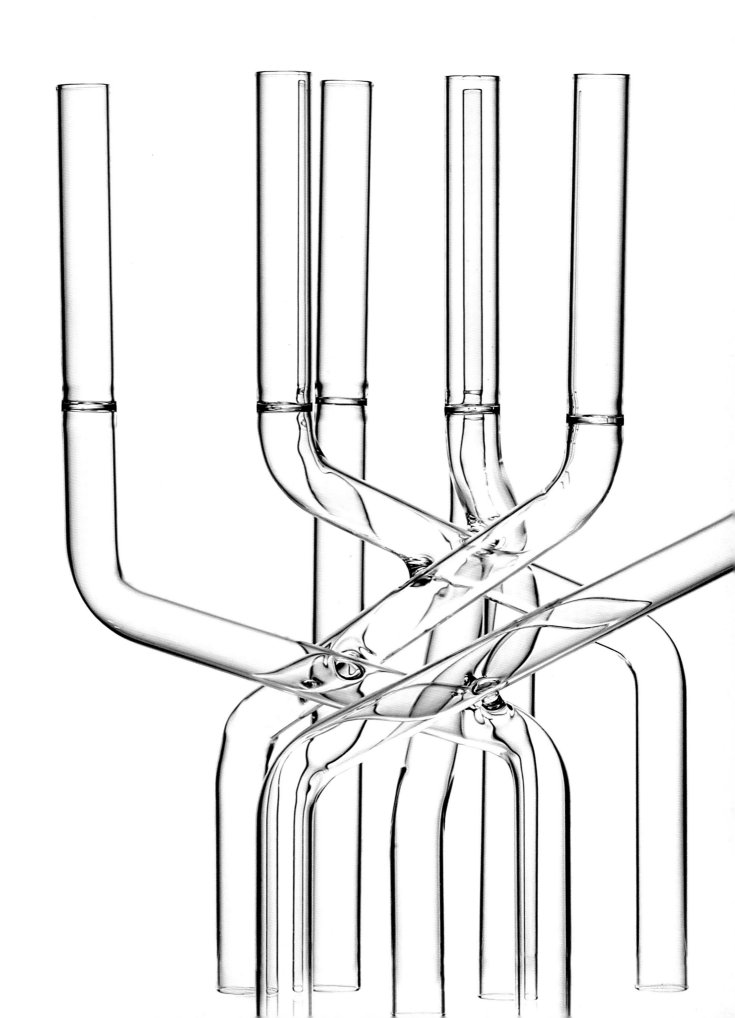

Bodum Pavina tumbler

The Pavina double-wall thermo tumbler is made from Borosilicate glass. Normally used in the fabrication of laboratory instruments, the material is strong and heat-resistant but weighs less than traditional glassware. Borosilicate substitutes boron oxide in place of soda and lime. The oxide acts as a glue, holding the silicate together, and due to the small size of the particles, the glass has a tighter structure, which results in increased strength. The same material has been used in Arik Levy's Mistic candleholder manufactured by Gaia & Gino (*see* page 171).

Double wall thermo tumblers, Pavina
Bodum AG/Bodum Design Group
Borosilicate glass
Diam: 6.4–9cm (2½–3½in)
Bodum/Bodum Design Group,
Switzerland
www.bodum.com

Double wall thermo glasses, Assam
Bodum/Bodum Design Group
Borosilicate glass
Diam: 6.8–9.5cm (2¾–3¾in)
Bodum/Bodum Design Group,
Switzerland
www.bodum.com

Drinking glasses and carafe,
Dance of the Glasses
Franz Maurer
Glass
Glasses: H: 7.5, 9.5 or 11.5cm
(3, 3¾ or 4½in); Diam: 9cm (3½in)
Carafe: H: 26.4cm (10⅜in),
Diam: 17cm (6¾in)
Raumgestalt GmbH, Germany
www.raumgestalt.net

Vases, Family Vases
Ineke Hans
Pewter, matt and polished
H: 14, 22 or 30cm
(5½, 8⅝ or 11⅞in)
W: 15cm (6in)
Tingieterij Leerdam,
The Netherlands
www.inekehans.com

Dish, Splash
Tim Parsons
Pewter
H: 6cm (2⅜in)
Diam: 20cm (7⅞in)
A.R. Wentworth (Sheffield), UK
www.wentworth-pewter.co.uk

Vases, Croce Grande (left) and
Fiore (right)
Karim Rashid
Ceramic
Croce Grande: H: 40.5cm (16in),
Diam: 22cm (8⅝in)
Fiore: H: 45cm (17¾in),
Diam: 22.5cm (8⅞in)
Flavia, Italia
www.flavia.it

Tableware, Nymphenburg Sketches
Hella Jongerius
Porcelain
Bowl with bird:
Diam: 22cm (8⅝in)
Bowl with fawn:
Diam: 22.5cm (8⅞in)
Limited-batch production
Nymphenburg Porcelain
Manufacturing, Germany
www.nymphenburg-porzellan.com

China vessels, Own Brand and
Own Brand Polka Editions
Marc Boase
Fine bone china
H: 20cm (7⅞in) (largest)
Diam: 8cm (3⅛in) (largest)
Limited-batch production
English Elegance, UK

Cup or vase, Random Cup
Marek Cecula
Porcelain, decal
H: 12cm (4¾in)
Diam: 11cm (4⅜in)
Limited-batch production
Modus Design, Poland
www.modusdesign.com

Kitchen knife,
Knife Programme IF4000
Industrial Facility: Sam Hecht,
Kim Colin, Ippei Matsumoto
Fully forged hi-carbon stainless
steel, melamine and
polyester moulding
L: 15–30cm (6–11¾in)
W: 2cm (¾in)
H: 2.5cm (1in)
Harrison Fisher, UK
www.harrison-fisher.co.uk

Glassware, Starck Darkside
Crystal Series
Philippe Starck
Onyx crystal, clear crystal
Various dimensions
Baccarat, France
www.baccarat.fr

Tableware, Biarritz and its Flowers
Marta Laudani and Marco Romanelli
in collaboration with
Laura de Ludicibus
Fine porcelained stoneware
Various dimensions
Laboratorio Pesaro, Italy
www.laboratoriopesaro.com

Vase, Vase of Phases
Dror Benshenrit
Porcelain
H: 21cm (8¼in), 26cm (10¼in)
or 30cm (11⅞in)
Rosenthal, Germany
www.rosenthal.de

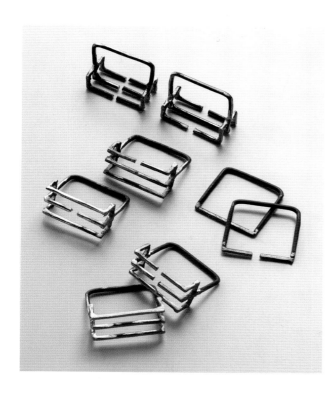

Napkin holder set, A-Shirin
Antonino Sciortino
Sterling silver
H: 2cm (¾in)
W: 5cm (2in)
L: 5cm (2in)
De Vecchi, Italy
www.devecchi.com

Fruit bowl, Baskettini (bottom)
Renaud Thiry
Steel, PTFE coating
Various dimensions
Limited-batch production
Flandesign, France

Fruit bowl, Meridian (Panier) (top)
Renaud Thiry
PVC, fabric
H: 16.5cm (6½in)
Diam: 33cm (13in)
Limited-batch production
Flandesign, France

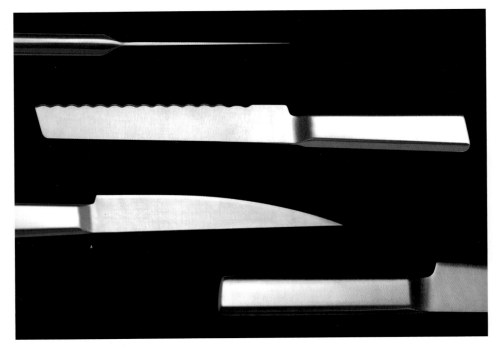

Containers for the table
and kitchen, Shuttle
Stefan Diez
Glass, plastic, porcelain
Various dimensions
Rosenthal, Germany
www.rosenthal.de

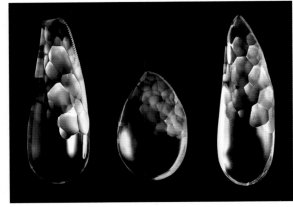

Kitchen knives,
Knife Programme IS3000
Sam Hecht
Carbon stainless steel
Various dimensions
Harrison Fisher, UK
www.harrison-fisher.co.uk

Plastic cutlery, De Luxe
Fabio Bortolani and
Donato Parruccini
Plastic
Fork: L: 21cm (8¼in)
Spoon: L: 21cm (8¼in)
Knife: L: 21.3cm (8⅜in)
Teaspoon: L: 14cm (5½in)
Pandora Design, Italy
www.pandoradesign.it

Knives, Glass Knives
Noa Bembibre
Glass, blown and carved
H: 4.5cm (1¾in)
W: 7cm (2¾in)
L: 18cm (7⅛in)
Limited-batch production
Skanno, Finland
www.skanno.fi

Tray with dragon carving, Dragon
Tord Boontje
Sterling silver
L: 45cm (17¾in)
W: 30cm (11⅜in)
H: 2cm (¾in)
Limited-batch production
De Vecchi, Italy
www.devecchi.com

Tableware, Showtime Collection
Jaime Hayón
Porcelain
Blue Napoleon: H: 59cm (23¼in)
Vulcano Grey: H: 42cm (16½in)
Pink Royale: H: 48cm (19in)
Oil Black: H: 33cm (13in)
Electric Yellow: H: 34cm (13⅜in)
Bd Ediciones de Diseño, Spain
www.bdbarcelona.com

Teaspoon, MP (Prime Minister)
Ed Annink, Ontwerpwerk
Polished stainless steel
W: 2.6cm (1in)
D: 1cm (³⁄₈in)
L: 11.5cm (4½in)
Ontwerpwerk, The Netherlands
www.ontwerpwerk.com

Candleholder, Von Erlach II
Marta Laudani and Marco
Romanelli in collaboration
with Marcello Pinzero
Polished aluminium casting
H: 2.7cm (1in)
W: 28cm (11in)
L: 40cm (15¾in)
Driade, Italy
www.driade.com

Vase, Huras Vase (opposite)
Andrea Branzi
Natural black ceramic 'bucchero',
with gold- or silver-plated
base and frame
H: 49cm (19¼in)
W: 34cm (13⅜in)
Limited-batch production
Design Gallery Milano, Italy
www.designgallerymilano.com

Bowl, Husque
Marc Harrison
Recycled macademia shell (80%)
and polymer
H: 8cm (3⅛in)
W: 18cm (7in)
L: 22cm (8⅝in)
One-off
Husque, Australia
www.husque.com

Tableware, Warbowl
Steve Mosley and Dominic Wilcox,
Mosleymeetswilcox
H: 10cm (4in)
Diam: 44cm (17⅜in)
www.mosleymeetswilcox.com

Front Design, Carbon Fibre Vases

Front Design's work is a million miles away from the modernist tradition we usually associate with Sweden. All their products share an emotional charge as well as a strong narrative. They are concerned with process and constantly question how objects are formed, harnessing the uncontrollable to arrive at often unexpected results. The vases illustrated here are formed by blowing glass within a sheath of knitted carbon fibre. The coating shapes the glass as it hardens.

Vases, Carbon Fibre Vases
Front Design
Carbon fibre, glass
H: 50cm or 45cm (19⅝ or 17¾in)
Prototypes
Front Design, Sweden
www.frontdesign.se

Fruit Bowl, Black Honey
Arik Levy
Epoxy
H: 11.4cm (4½in)
L: 39.5cm (15½in)
W: 40.3 cm (15⅞in)
Materialise.MGX, Belgium
www.materialise-mgx.com

Bowls, Eggshells
Pieke Bergmans
Ceramic, translucent metallic
coloured lacquer
Various sizes: Diam: 12–35cm
(4³⁄₄–13³⁄₄in)
Prototype
www.piekebergmans.com

Vase, Memory Pot
Marta Daza Fernández
Polyethylene
H: 44cm (17³⁄₈in)
W: 53cm (20⁷⁄₈in)
L: 49cm (19¹⁄₄in)
Serralunga, Italy
www.serralunga.com

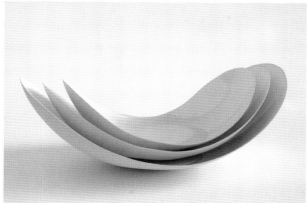

Ceramics, Filmpaper Ceramics
Vittorio Passaro
Ceramics, paper
H: 13cm (5¹⁄₈in)
W: 13cm (5¹⁄₈in)
L: 13 cm (5¹⁄₈in)
Diam: 13cm (5¹⁄₈in)
Prototype

Photo and memory container, Memory
Box (models one and two) Sebastian
Bergne
Extra clear crystal bonded glass
One: H: 38cm (15in),
W: 26.5cm (10³/₈in),
D: 8.3cm (3¼in)
Two: H: 20.5cm (8in),
W: 36.5cm (14³/₈in),
D: 8.3cm (3¼in)
RSVP, Italy
www.r-s-v-p.it

Dinner party tray, Nibbles
Steven Koch
Recycled stainless steel
H: 0.8cm (³/₈in)
W: 10cm (4in)
L: 25cm (9⁷/₈in)
Limited-batch production
Play Design, UK
www.iplaydesign.com

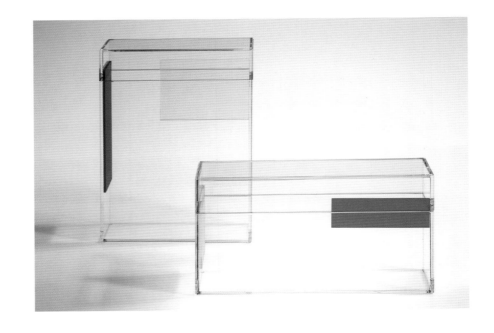

Vase, Lounge Vase
James Irvine
Cromargan
(stainless steel 18/10)
H: 28cm (11in) or 18cm (7in)
Diam: 11cm (4³/₄in) or 9cm
(3½in)
WMF Wurttembergische
Metallwaren Fabrik, Germany
www.wmf.de

Tableware, Incognito
Khashayar Naimanan
Hand-thrown porcelain
Various dimensions
Limited-batch production
Nymphenburg Porcelain
Manufacturing, Germany
www.nymphenburg-porzellan.com

Teapot, Mirza
Catherine Lévy and Sigolene
Prébois, Tsé and Tsé Associates
Porcelain, aluminium, rubber,
wood, fabric
H: 14cm (5½in)
L: 25cm (9⅞in)
Diam: 10cm (3⅞in)
Limited-batch production
Tsé and Tsé Associates, France
www.tse-tse.com

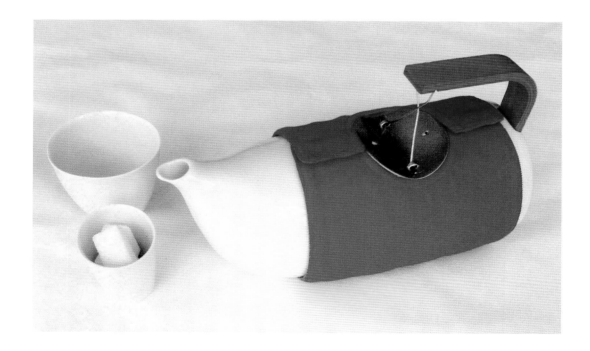

Vases, The Other Side of
the Ceramics collection
Tord Boontje
Ceramics
Various dimensions
Moroso, Italy
www.moroso.it

Vase, Ikea PS Jonsberg
Hella Jongerius
Stoneware, feldspar porcelain
H: 34cm (13⅜in)
Diam: 30cm (11⅞in)
IKEA of Sweden, Sweden
www.ikea.com

Candelabrum, Giants:
Herkimer (left) and Helidor (right)
Arik Levy
Clear crystal
H: 17.5cm (6⅞in) or 20cm (7⅞in)
Gaia & Gino, Turkey
www.gaiaandgino.com

Vika Mitrichenka, Grandmother's Treasures

Vika Mitrichenka's nostalgic porcelain service
was commissioned by The Frozen Fountain shop.
Inspired by memories of her Russian grandmother,
who was so proud of her fine china tea service
that she would never throw away a single item
even if broken, Mitrichenka has fired an entirely
new series of cups, saucers and teapots, which
deceptively appear to have been made by gluing
together old, found bits and pieces. Playing on our
notion of fragmentary recollections, imagination
and dreams, she has assembled a service that looks
as if it has parts missing or has been stuck together
incorrectly, but which in fact forms a cohesive
whole, albeit with diverse styles and characters.

Dinner service,
Grandmother's Treasures
Vika Mitrichenka
Porcelain
Various dimensions
Vika Mitrichenka's studio,
commissioned by
The Frozen Fountain Shop,
The Netherlands
www.frozenfountain.nl

Tableware, KU Table Set
Toyo Ito
White porcelain
Various dimensions
Alessi, Italy
www.alessi.com

Dishes, Lux-it
Alexander Estadieu
Porcelain, metal, plastic
H: 8–12cm (3⅛–4¾in)
Diam: 15–25cm (6–9⅞)
Prototype

Tableware, Biscuit Collection
Studio Job
Porcelain
Various dimensions
Royal Tichelaar Makkum,
The Netherlands
www.tichelaar.nl

Tableware, Creamer Collection No 2, 4 and 17
5.5 Designers
Limoges porcelain
No 17: H: 12.5cm (4⅞in), Diam: 5.5cm (2⅛in)
No 2: H: 4cm (1⅝in), Diam: 4cm (1⅝in)
No 4: H: 8.5cm (3⅜in), Diam: 4cm (1⅝in)
Prototype
Foundation Bernardaud, France
www.bernardaud.fr

Tableware, Porcelain Vessels
Mieke Everaet
Coloured porcelain
Various dimensions
One-off
Studio Mieke Everaet, Belgium
www.pulsceramics.com

Tableware, Arita Houen
Motomi Kawakami
Porcelain
Various dimensions
Kihara, Japan
www.aritahouen.jp

Tableware, Pitcher
Motomi Kawakami
Glass, silicon
H: 13cm (5⅛in)
Diam: 7cm (2¾in)
Maruhachi Seichajo, Japan
www.kagaboucha.co.jp

Kitchen tongs, Tong Tool
Gary Allson
Cherry, sycamore, maple,
variety of fruit woods
Various dimensions
Limited-batch production
www.hiddenartcornwall.co.uk

Set of dishes, glasses and
cutlery for children,
Kids' Stuff
Alfredo Häberli
Glass, wood, metal, plastic
Various dimensions
Iittala Oy, Finland
www.iittala.com

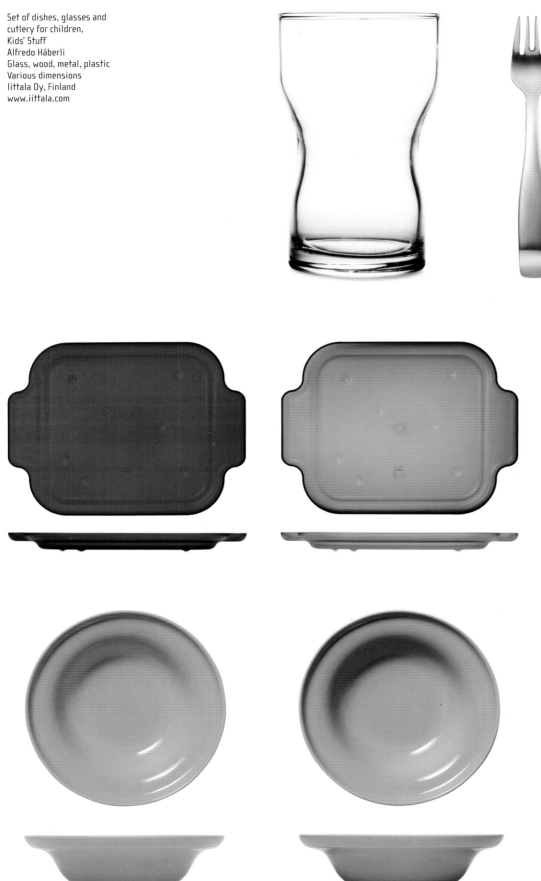

Tableware, My China!
Michael Sieger
Porcelain, gold
Various dimensions
Limited-batch production
Fürstenberg, Germany
www.furstenberg-porzellan.com

textiles

60

60° was

A

ot iron

Surface Patterns, (left to right):
Orange boxes, EC2 Floral Stamps,
ECI Floral Pink
Debbie Jane Buchan
L: 353cm (139½in)
W: 518cm (204in)
One-off
debbiejanemagee@hotmail.com

Rug, Disorder
Adrien Rovero
100% wool
L: 240cm (94½in)
W:175cm (69in)
Tisca Tiara, Switzerland
www.tisca.ch

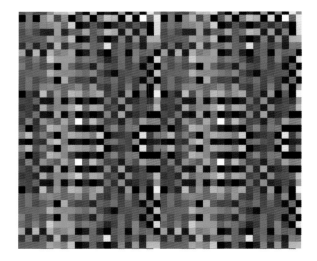

 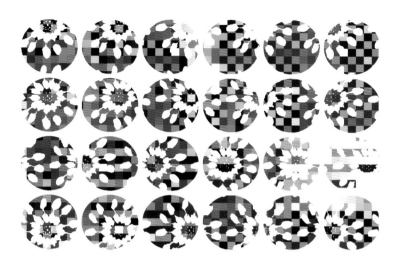

Rug, Rose Lace
Kiki van Eijk
100% wool, hand-tufted
H: 1.5cm (⁵⁄₈in)
W: 170cm (67in)
L: 240cm (94½in)
Limited-batch production
Danish Carpets, The Netherlands
www.kikiworld.nl

Passiflora rug, Botanica
Rosita Missoni
100% wool
Diam: 110 cm (43³/₈in)
Limited-batch production
T&J Vestor, Italy
www.missonihome.it

Wallpaper/textile, Mind the Gap
Ane Lykke
Transparent acrylic and
translucent film
H: 350cm (137³/₄in)
W: 550cm (216¹/₂in)
D: 14cm (5¹/₂in)
One-off

Ane Lykke, Mind The Gap
'I want to find new ways of affecting the
perception of space,' says Ane Lykke. Mind the Gap
is a 3D installation: a wall formed by two layers
of hexagonal plastic boxes standing 14cm (5¹/₂in)
apart is patterned by a number of red lines of
varying widths. The result is what physicists refer
to as interference patterning. As people move
across a room, the 'wallpaper' appears to change,
the optical play challenging our expectations and
appealing to our senses.

Installation, Rhythm of Shadow
Maria Luisa Brighenti
Porcelain tile, opaline glass, steel
H: 300cm (118¹⁄₂in)
W: 400cm (157¹⁄₈in)
D: 600cm (236¹⁄₄in)
One-off
Rex Ceramiche Artistiche, Italy
www.rex-cerart.it

Wall and floor tiles, Ma.De
Cocos Tortora
Iris Ceramica Style and Design
Research Laboratory
Gres porcelain
W: 60cm (23⁵⁄₈in)
L: 60cm (23⁵⁄₈in)
Iris Ceramica, Italy
www.irisceramica.com

Décor on sheet glass, Tree
Marc Krusin
Acid-etched,
back-painted glass
H: 200cm (79in)
W: 300cm (118½in)
Various thicknesses
Omnidecor, Italy
www.omnidecor.net

Wallpaper (magnet receptive)
and decorative magnets,
MagScapes
Patricia Adler
Specially treated wallpaper
and flat-sheet magnets
W: 52cm (20½in)
L: 500cm (196⅞in)
Pepper-mint (UK), UK
www.pepper-mint.com

Patricia Adler, MagScapes wallpaper
Using sets of matching magnetic motifs and a
specially treated wallpaper, MagScapes allows
the user to create his or her own designs and
interact playfully with the environment without the
fear of making mistakes. 'Play is the gateway into
creativity. Beautiful paper and motifs distinctly
decorate any room and transform its possibilities,'
says Patricia Adler, design director of Pepper-mint,
which patented the concept.

Concrete Blond, Walled Paper

As yet the technique behind Eric Barrett's concrete walled paper is top secret, but the result caused the biggest stir not only at Designers' Block but at the whole of London's 2005 '100% Design'. Produced to commission, it is available in different sizes, colours and bespoke patterns and is cast on site in continuous lengths. This cladding system can be used for both interior and exterior applications.

Building cladding system (internal and external),
Walled Paper
Eric Barrett, Concrete Blond
Cast concrete
Various dimensions
Concrete Blond, UK
www.concrete-blond.com

Wallpaper, Birds
Ed Annink, Ontwerpwerk
Full-colour printed paper
Various dimensions
Designwall, The Netherlands
www.designwall.nl

Lene Toni Kjeld, Walhalla wallpaper

Inspired by the kaleidoscopic and mismatching surface patterns seen in a photograph of a Russian living room, Kjeld's graduation project is concerned with designing a wall covering that gives a room a unique look while creating different functional zones. Colours and patterns based on geometry or floral motifs collide and merge by way of a special transition roll, which links one idea to the next. Although it is still a prototype, the idea is to customize the wallpaper for individual applications.

Wallpaper, Walhalla
Lene Toni Kjeld
Non-woven wallpaper
W: 53cm (20⅞in)
L: 104m (341ft)
Prototype
Fiona, Denmark
www.fiona-walldesign.com

Fabric tiles,
Stockholm Kvadrat Showroom
Ronan and Erwan Bouroullec
Foamcore laminated with fabric
H: 24cm (9½in)
W: 65cm (25⅝in)
D: 1.5cm (6in)
Kvadrat, Denmark
www.kvadrat.dk

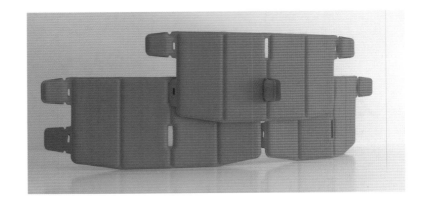

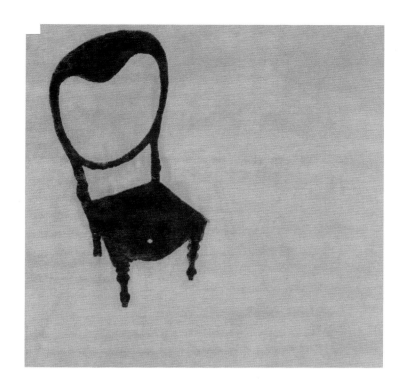

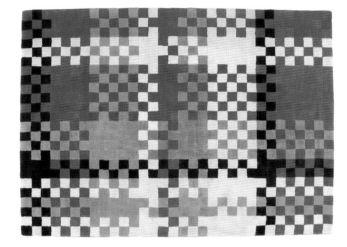

Rug, Chair Silhouette
Rosemary Hallgarten
Wool and silk
W: 71cm (28in)
L: 71cm (28in)
Limited-batch production
Rosemary Hallgarten, USA
www.rosemaryhallgarten.com

Rug, Tartan
Michael Sodeau
100% wool
W: 180cm (71in)
L: 240cm (94¹/₂in)
Modus, UK
www.modusfurniture.co.uk

Textile handprint,
Bauhaus on Speed
Vibeke Rohland
Pigment, cotton
H: 210cm (82⁵/₈in)
W: 150cm (59in)
One-off

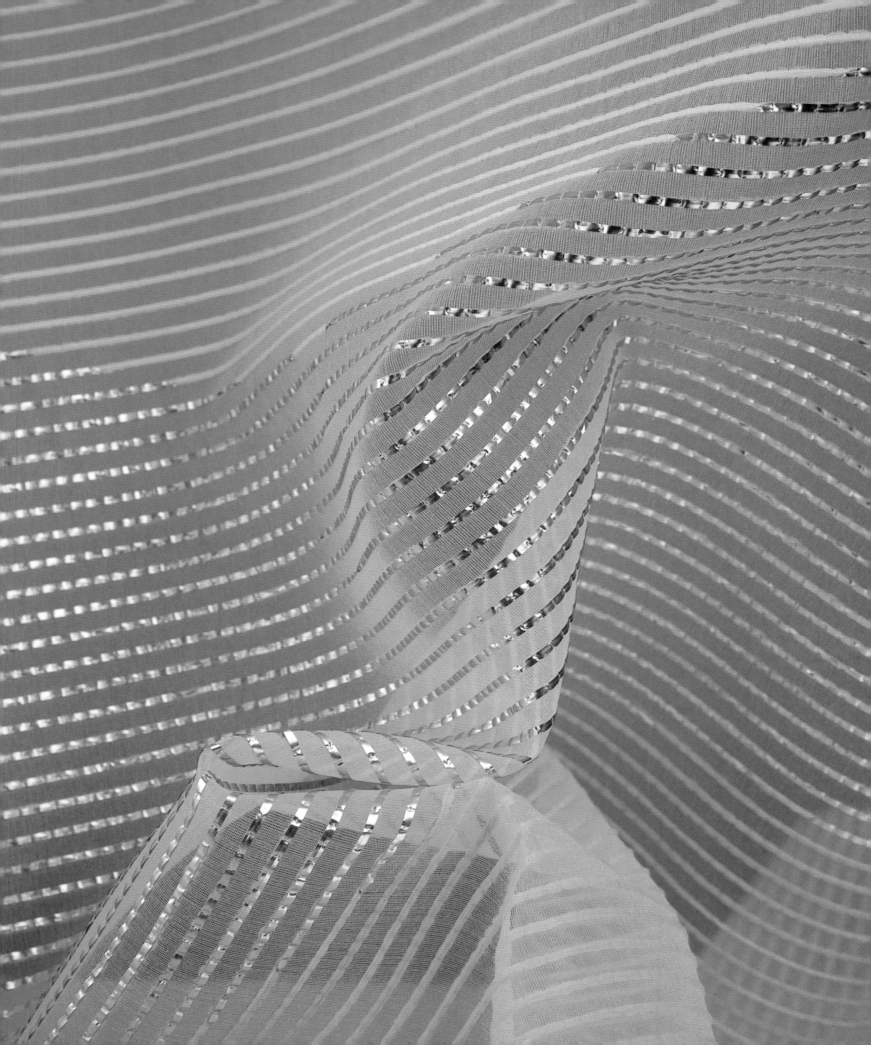

Jasper van Grootel, Cover Tiles

The concept developed from the desire to express a bathroom wall's surface by what's hidden within. The pipes and water taps, which are normally concealed, are made visible, becoming part of the ceramic tile wall-covering, revealing the raw, technical system of water lines, their connections and how they work. The series consists of five different 15 x 15cm (5⅞ x 5⅞in) tiles. The package comes with a manual for cutting and bending the pipes to the right length so the tiles can be used in a custom-made fashion. The producer, Cor Unum, makes handmade ceramics for designers.

Tiles, Cover Tiles
Jasper van Grootel,
Studio JSPR
Ceramic
L: 15cm (5⅞in)
H: 15cm (5⅞in)
Cor Unum, The Netherlands
www.studiojspr.nl
www.corunum.nl

Decoration, Tiburon
Design-Team Nya Nordiska
65% polyester, 30% polyamid,
5% pes-metal
W: approx. 165cm (65in)
Nya Nordiska Textiles, Germany
www.nya.com

Wall or floor tile, Crosstile
Mark Gabbertas
6mm float glass
H: 17.5cm (6⅞in)
W: 17.5cm (6⅞in)
i glass, UK
www.iglass.co.uk

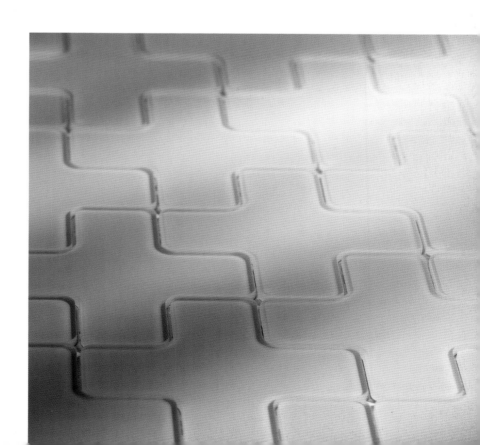

LiTraCon, light-transmitting concrete

Adding to the plethora of translucent materials being used by architects today, LiTraCon (the name of both product and company) is the brainchild of Hungarian architect Aron Losonczi. Light-transmitting concrete is formed by piercing concrete with thousands of glass and plastic fibres, which allow light to travel through and give the concrete a deceptive soft and spongy texture. Losonczi's business partner Andreas Bittis maintains that the fibres permit light, whether natural or artificial, to penetrate through the blocks for up to 20 metres (66ft) with no loss of intensity. Silhouettes are clearly visible through the wall and even colours remain unaffected. The material is available in a range of shapes and sizes from paving stones to exterior panelling. Its properties are the same as normal concrete.

Pre-cast translucent concrete elements (plates, blocks, stones),
LiTraCon light-transmitting concrete
Andreas Bittis with Aron Losonczi
Cement, sand, water, optical fibres, additives
Various dimensions
Limited-batch production
Andreas Bittis, Germany
www.litracon.com

Richardson and Page, Fly Tip wallpaper

The designs of Harry Richardson and Clare Page are almost always a direct response to their surroundings. They live and work in southeast London, an area plagued by fly-tipping (illegal dumping). The Fly Tip wallpaper combines images of discarded objects they have found in the street with their passion for panoramic wall-coverings from nineteenth-century France.

Wanting to make an item of beauty and luxury from the rubbish of others, they photographed items that reference all of the things that usually make up the stuff we throw away – food, clothing, old electronics and modes of communication – mixing them with elements of desire such as a doll and wedding ring as well as traditional wallpaper motifs such as flowers and birds. Cellphones and circuit boards play an important role, as it was important for the designers to fix the imagery very much of our times in order to create something that will eventually look out-of-date.

The wallpaper is screen-printed by hand using more than twice as many colours as mass-manufactured paper, which makes it very expensive and impossible to produce in a repeat pattern. Designed to be used in conjunction with a matching affordable paper, the garland of detritus floats against an idyllic sky-blue background and is intended to be viewed in a solitary row.

Wallpaper, Fly Tip
Harry Richardson and Clare Page
Paper, hand screen-printed by Cole and Son
W: 52cm (20½in)
L: 10m (33ft)
Limited-batch production, Committee, UK
www.gallop.co.uk
www.cole-and-son.com

rAndom International PixelRoller

The PixelRoller is a collaborative project from two ex-Royal College of Art graduates, Stuart Wood, who studied interaction design, and Florian Ortkrass, who studied design products. This innovative creative tool prints pixels exported directly from the internet, cellphone or camcorder onto any surface. The content is applied in continuous strokes by the user with a hand-held 'printer', which is based around the ergonomics of a paint roller. It can be loaded with paint, ink, water or even cement. The concept behind the product is to offer the flexibility of manual printing with accurate reproduction of digital information, allowing the user to influence the output in real time.

Communication tool, rapid-response printing device,
PixelRoller
rAndom International
Various materials
H: 15cm (5⅞in)
W: 40cm (15¾in)
L: 15cm (5⅞in)
Prototype
rAndom International, UK
www.random-international.com

Mirrors, Mirror Virus
Pieke Bergmans
Mirrored acrylate
Various sizes: H: 10–150cm (4–59in)
Limited-batch production
Studio Design Virus, The Netherlands
www.piekebergmans.com

Broersen and Lukács, Black Light wallpaper
Persijn Broersen and Margit Lukács' Black Light
wallpaper is an essay in the beauty of chaos.
Based on 3,000 photographs of Spanish flora that
the artist duo took while on holiday, flowers, stems,
leaves and seeds blossom out of control and spread
across a jet-black background. A once classical
pattern degenerates into random disorder and
finally disintegrates. 'Patterns today are either
getting too complex or falling apart, and we see

no way to influence this process. In designing
this wall covering, we tried to translate that
gloomy feeling into something as hypnotizing and
attractive as it is alienating.'

Wallpaper, Black Light
Persijn Broersen and Margit Lukács
Paper
Various dimensions
Site-specific artwork
www.pmpmpm.com

Rug, Frori
Marta Laudani and Marco Romanelli
in collaboration with
Laura de Ludicibus
Hand-woven wool and cotton
L: 200cm (79in)
W: 140cm (55in)
Limited-batch production
Imago Mundi for SardegnaLab, Italy
www.Imagomundidesign.com

Erotic wallpaper, Oriental Orchid (below)
Paul Simmons
Ink, paper
W: 52cm (20½in)
Timorous Beasties, UK
www.timorousbeasties.com

Textiles, Organic Non-woven (below right)
Sylvia Döhler
Pine needles and fibres
L: 100cm (39in)
W: 30cm (11¾in)
Sylvia Döhler, Germany
Limited-batch production
sdoehler@mac.com

Sylvia Döhler, Organic Non-woven

Non-wovens are normally used in technical applications. Döhler works with natural materials to 'bring back a sense of nature to our urban lives'. The fabric was created by training fine, high-speed jets of water onto a web of fibres, which has the effect of rearranging and entangling them into a random and interlocking spun-lace web. Pine needles or leaves have been worked in between the layers during the production process, adding to the unusual texture and dying the textile with earthy biological colours: a perfect symbiosis of the natural and high-tech.

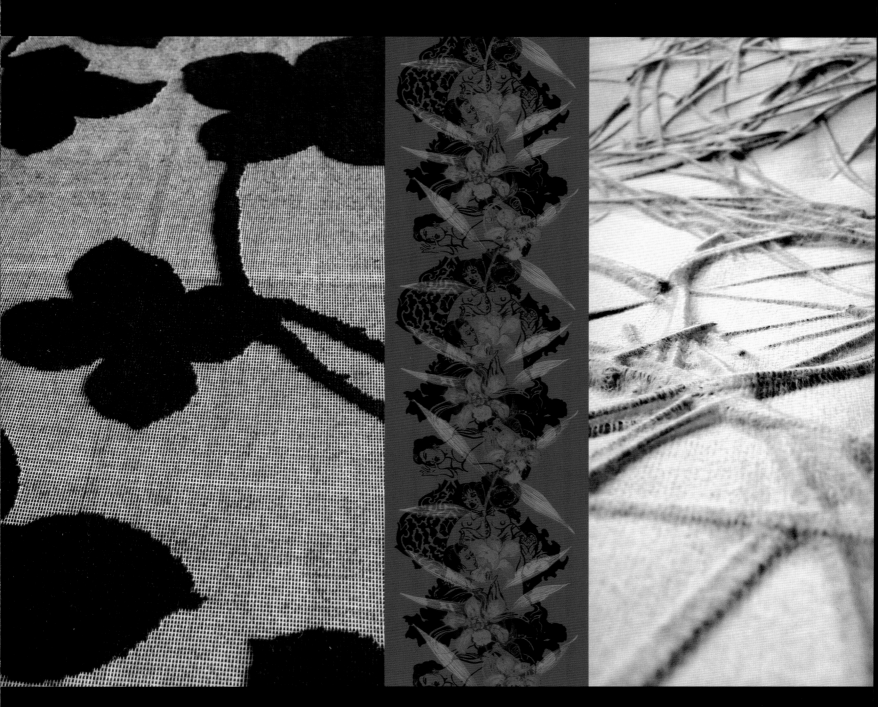

Dahlia rug, Avalon
Rosita Missoni
100% wool
Diam: 220cm (86⅝in)
Limited-batch production
T&J Vestor, Italy
www.missonihome.it

Inlay decoration on cabinet
Perished Collection
Job Smeets and Nynke Tynagel,
Studio Job
Hardened tropical woods,
decorated with inlays
Studio Job, Belgium
www.studiojob.nl

Timorous Beasties, London and Glasgow toiles

There has been a long tradition of politicizing drawing rooms by the use of artfully designed wallpapers. As far back as the eighteenth century there were patterns of the Bastille being razed to the ground by French revolutionaries, a Britannia made to suffer at the hands of American patriots, images of Napoleon's campaigns and Hellenic Greeks battling against marauding Ottoman barbarians. Even in the White House, Jackie Kennedy hung Revolution scenes, which were covered over by the Clintons and revealed once more by the Bushes.

Timorous Beasties is returning to the tradition with its Glasgow and London Toiles. Alistair McAuley and Paul Simmons founded their own studio in 1990,

and today run one of the few companies that both design and produce its own fabrics under one roof. Described as 'William Morris on Acid', their innovative patterns depicting contemporary images, which are silkscreen-printed onto traditional textiles and wallpapers, have won them international acclaim and a nomination for Designer of the Year from the Design Museum in 2005.

The toiles are a modern take on the eighteenth-century version, which was produced in the Jouy factory and which traditionally depicted scenes of pastoral idyll set in magnificent vistas. The patterns on the modern-day fabrics may, at first glance, appear benign and bucolic, but closer examination reveals that something political is going on, some-

thing subversive. Timorous Beasties argues that the imagery of the eighteenth century wasn't actually that innocent, concentrating as it did on scenes that were then contemporary: some wallpapers showed the Jouy factory itself, while others presented rural scenes of workers relaxing, dancing and womanizing.

Bringing the French countryside into inner-city Glasgow, the Beastie boys maintain that they haven't changed the concept much: 'a glass of wine became a can of super lager, a pipe a rollie, and an old man sitting on a stool in a rural scene became a tramp on a park bench'. Also borrowing from morality paintings of the time, graphic images take on symbolic meaning. A junkie shoots up in a graveyard (the famous Glaswegian Necropolis much favoured by drug

abusers), the moral being that if you inject heroin then you are likely to end up in the graveyard – permanently; a young boy pisses against a tree, while an old tramp takes a swig of lager – if you start misbehaving too early then you will end up homeless and drunk; Glasgow University towers above it all like a fairy-tale castle, while Foster's Armadillo Building represents the changes that are happening along the Clyde.

Similarly the London toile depicts disaffected youths in hoodies, alienated office workers, pigeons, sellers of *The Big Issue* magazine, all going about their business in the shadows of East End tower blocks and London landmarks.

McAuley and Simmons have a love of traditional elements like academic drawings, use of complicated repeats, and the hand-printed quality of inks, and call their work 'modern tradition'. 'We do love some of the traditional designs from the past, but it's great fun to give them a new angle, to make them speak to us in the present.'

Fabric, London Toile
Timorous Beasties
Linen
W: 135cm (53in)
Timorous Beasties, UK
www.timorousbeasties.com

Fabric, Glasgow Toile
Timorous Beasties
Linen
W: 136cm (53½in)
Timorous Beasties, UK
www.timorousbeasties.com

Sharon Walsh

Sharon Walsh's designs are inspired by the nostalgic quality found in unwanted objects. She reuses household fabrics to recreate innovative and one-off pieces for both fashion and interiors. For example, by applying a combination of deconstructive techniques to bind layers of fabric, an old pair of curtains can be reincarnated into a sophisticated fashion garment or bring life back into an old chair.

Textile, Floral Spray
Sharon Walsh and Andra Nelki
Flock, silk velvet, vintage curtains
Various dimensions
One-off
Sharon Walsh, UK
www.sharonwalsh.co.uk

Textile, Black Lace Panel
Lauren Ruth Moriarty
Laser-cut neoprene
W: 100cm (39³⁄₈in)
L: 300cm (118¹⁄₈in)
Limited-batch production
Lauren Moriarty, UK
www.laurenmoriarty.co.uk

3D Textiles, Noodle Block Cube
Lauren Ruth Moriarty
Laser-cut neoprene
H: 60cm (23⁵⁄₈in)
W: 60cm (23⁵⁄₈in)
L: 60cm (23⁵⁄₈in)
One-off
Lauren Moriarty, UK
www.laurenmoriarty.co.uk

Tiles, Illusion
Erika Lövqvist
Ceramic
H: 20cm (7⁷⁄₈in)
W: 20cm (7⁷⁄₈in)
Prototype
Erika Lövqvist Design, Sweden
www.erikalovqvist.se

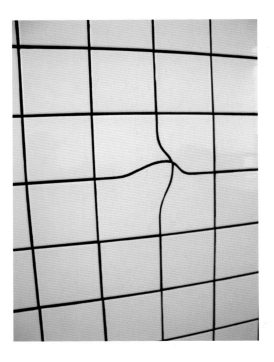

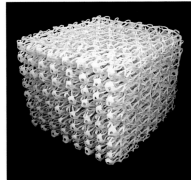

Textile (casement), Cat's Cradle
Donghia Design Studio
100% polyester
W: 150cm (59in)
Limited batch production
Donghia, USA
www.donghia.com

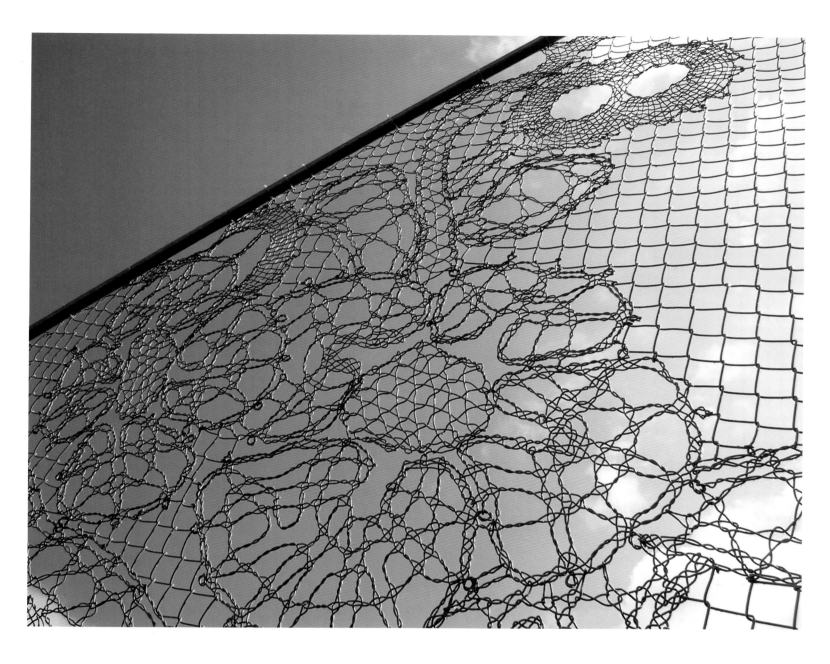

Fence, LaceFence
Joep Verhoeven
Option 1: completely galvanized
Option 2: metal wire with
plastic coating
Various dimensions
Demakersvan, The Netherlands
www.demakersvan.com

Textile, Hana
Yoshiki Hishinuma
Inkjet print and embroidery
on leather
L: 25cm (9⅞in)
W: 30cm (11¾in)
Limited-batch production
Yoshiki Hishinuma Co., Japan
www.yoshikihishinuma.co.jp

Carpet, Dialogue
Ane Lykke
Paper
H: 45cm (17¾in)
W: 150cm (59in)
L: 300cm (118⅛in)
Prototype

Ane Lykke, Dialogue carpet

Dialogue challenges the concept and traditional
function of a carpet: it can be shaped and modelled
in endless configurations. Users are invited to
interact and play, creating a fresh way to experience
space and connect one with another.

Tiles (digitally manipulated
photograph), Jafleur
Dominic Crinson
Printed ceramic wall tiles
H: 20cm (7⅞in)
W: 20 cm (7⅞in)
Digitile: Dominic Crinson, UK
www.crinson.com

Rug, Roses
Nani Marquina
100% wool, dyed felt and cotton
W: 170cm (67in)
L: 240cm (94½in)
Nanimarquina, Spain
www.nanimarquina.com

Metal weaves (to be used for
interior decoration/ curtains), Titan
Création Baumann Design Team
70% metal, 30% polyester
W: 170cm (67in)
Création Baumann,
Weavers & Dyers, Switzerland
www.creationbaumann.com

Rug, Knot
Ulf Moritz
Polypropylene
H: 1.5cm (⅝in)
Danskina, The Netherlands
www.danskina.com

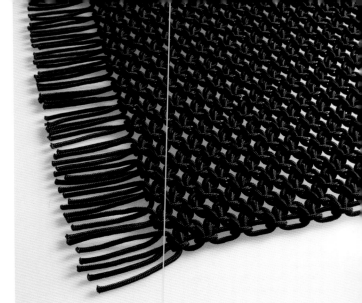

Rug, Dune
Ulf Moritz
100% pure new wool
Pile: H: 1.8cm (⅝in)
Custom made
Danskina, The Netherlands
www.danskina.com

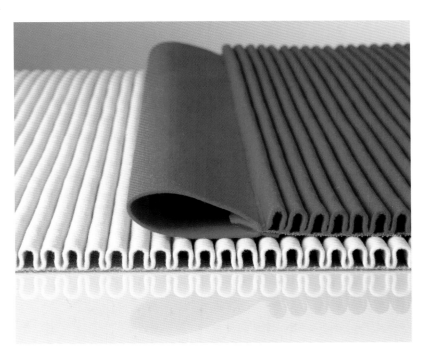

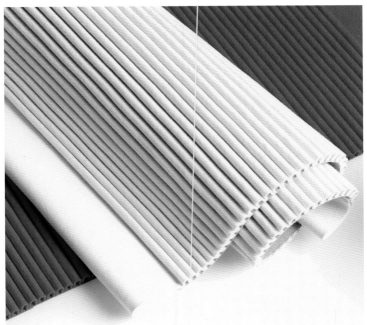

Rug, Paris Market
Kate Blee
100% hand-spun wool
L: 220cm (87in)
W: 150cm (59in)
Limited-batch production
Christopher Farr
Handmade Rugs, UK
www.christopherfarr.com

Rug, Salim
Alfredo Häberli
Wool felt
W: 160cm (63in)
L: 320cm (126in)
Ruckstuhl, Switzerland
www.ruckstuhl.com

Tile, Sculpture
Massimiliano Fuksas
Porcelain tile
W: 20cm (7⅞in)
L: 40cm (15¾in)
Limited batch production
Cerdomus Ceramiche, Italy
www.cerdomus.com

Tablecloth, double-sided;
Cinderella
Ineke Hans
Cotton, acrylic, lurex, viscose
W: 160cm (63in)
L: 60, 120, 180, 240 or 300cm
(23 5/8, 47 1/4, 70 7/8, 94 1/2,
118 1/2in)
National Textile Museum
Holland, The Netherlands
www.textilemuseum.nl

Laptop bag, Plissee Onepiece
Mary-Ann Williams
A single piece of felt (100% pure
wool), anodized aluminium
Available for 38cm (15in), 43cm
(17in) and 48cm (19in) laptops
Limited-batch production
Illu Stration, Germany
www.illu-stration.com

Textile, Les frites de B.I.C.
In-house design B.I.C.-Carpets
100 % wool (felted)
W: max. 400cm (157 1/2in)
L: max. 2000cm (787 3/8in)
H: 2.2cm (7/8in)
B.I.C.- Carpets from
Belgium
www.bic-carpets.be

Aleksandra Gaca, Weavepoint fabrics

Aleksandra Gaca's 3D fabrics are created on a detailed software program, which is offering textile designers the opportunity of creating seemingly endless combinations of yarns and weaving techniques. Weavepoint is a computer-controlled loom system. Through extensive experimentation, Gaca mixes a variety of patterns using single or double weaves with a myriad of yarns, merging mohair, cottons and man-made fibres – the soft, coarse, thick and thin. The result is a highly varied textile collection. Weavepoint also allows adjustments to be made during the process. Gaca can alter not only the patterns she has conceived but also the colourways and the choice between single and double weaves. This, she maintains, is an important feature because it is not always easy to tell from the image on the computer monitor whether her combinations will result in an attractive structure best suited to the final use of the fabric.

3D airflow upholster fabric,
Equinox XL and Meteor L
Aleksandra Gaca
Polyester
W: 140cm (55⅛in)
Hybrids + Fusion, The Netherlands
www.hybridsandfusion.com

3D weave, Slumber
Aleksandra Gaca
Wool, cotton, elastic
W: 120cm (47in)
Limited-batch production
TextileLAB, The Netherlands
Textile Museum, The Netherlands
www.textilemuseum.nl

Rug, Terrain
Louisa Vilde
Pure wool felt
W: 155cm (61in)
L: 198cm (78in)
Limited-batch production
Vilde Form
www.vildeform.com

Wall covering, Tibet
Claudy Jongstra
Merino, yak, Drenthe Heathe wool,
cotton gauze
Various dimensions
Limited-batch production
Studio Claudy Jongstra,
The Netherlands
www.claudyjongstra.com

Wall tiling, insulating tiling,
decorative surfaces;
Hard Coated Foam Tiles
Tjeerd Veenhoven
Polystyrene, polyester/cotton
lace (or metal mesh or any other
open fabric)
Various dimensions
Tjeerd Veenhoven, the Netherlands
www.tjeerdveenhoven.com

Rug, Bicicleta
Nani Marquina and Ariadna Miquel
100% recycled rubber
W: 170cm (67in)
L: 240cm (94½in)
Nanimarquina, Spain
www.nanimarquina.com

Textile, Multi-coloured
Modular Textile
Galya Rosenfeld
Reclaimed ultrasuede fabric
H: 302.3cm (119in)
L: 91.4cm (36in)
D: 2.5cm (1in)
One-off
Galya Rosenfeld, USA
www.galyarosenfeld.com

Textile, Camo
Yoshiki Hishinuma
100% polyester
W: 100cm (39⅜in)
L: 150cm (59in)
Limited-batch production
Yoshiki Hishinuma Co., Japan
www.yoshikihishinuma.co.jp

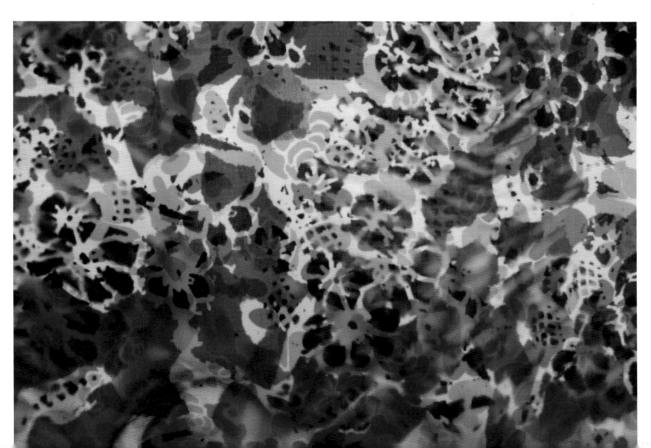

biographies

3PART was founded in 1998 by industrial designers Simon Skafdrup and Henning Therkelsen and has grown into one of Denmark's biggest design companies. The name '3PART' is derived from the company's mission 'to be the link between manufacturer and customer' through an integrated collaboration between inventors, engineers, industrial and graphic designers and communication designers. 3PART has become well known for its product designs for the elderly and disabled and has won numerous design awards. 110

5.5 Designers is an industrial-design research and consultation agency, which was founded in 2003 by Vincent Baranger, Jean-Sebastien Blanc, Anthony Lebossé and Claire Renard. Since first claiming attention in 2003 with their project 'Réanim – the Medicine of Objects', 5.5 Designers has cooperated with brands such as Arc, Ozé, Bernardaud and Galeries Lafayette. 194

Patricia Adler, designer–inventor, was born in Vienna, Austria, and studied furniture and product design at the Rhode Island School of Design in Providence and at Kingston University, London. In 1999 she founded the Pepper-mint (UK) design studio in London, and began to develop ideas for furniture and lighting. Her debut product, the Qube, was launched at the 'Salone Satellite' at the Milan Furniture Fair, and has since been sold by stockists in Europe, North America, Asia and Australia, including The Conran Shop, Purves & Purves, and The Apartment, New York. She has designed storage systems, lighting and window displays, and also commercial and private interiors, and her clients include Bloomberg, the BBC, Selfridges and Foster & Partners. 206

Harry Allen studied industrial design at the Pratt Institute in New York. He later opened his own studio and developed a line of furniture called Living Systems, which was manufactured for clients such as Habitat, Dune and Magis. He has also become known for his interior design, which includes a number of projects in New York, such as the Moss Gallery in Soho and offices for the Guggenheim Museum, and also retail interiors in Japan and Taiwan as well as residential interiors. Additionally, Allen produces product designs for a diverse group of clients, including Steuben, Aveda, Target, DuPont, Swarovski, George Kovacs and IKEA. His work has received a number of awards, has been shown in various exhibitions and is in the permanent collection of the Denver Museum of Art, Colorado. 99

Gary Allson is currently based in Falmouth, England. His research and designs are inspired by observations of daily life. He intentionally maintains simplicity in the materials and techniques he employs in order to emphasize the idea itself. 195

Ed Annink, born in 1956, is a partner in Ontwerpwerk design studio in The Netherlands. He studied in the department of furniture design and interior architecture at the Royal Academy for Art and Design, The Hague, and was the initiator of FunLab, the MA programme in experience and scenario design at the Design Academy Eindhoven. His work for Droog Design has been exhibited and published, and he has recently written a book entitled *Bright Minds, Beautiful Ideas* in collaboration with Ineke Schwartz. Annink organizes and leads many workshops, lectures, symposia and exhibitions. 180–1, 206–7

Apple Design Team, led by Jonathan Ive, grabbed the world's attention in 1998 with the release of the first iMac. More than two million iMacs were sold in the first year, and *BusinessWeek* cited its design as 'one of the century's lasting images'. The iMac went on to win many design competitions, including Best of Category in *ID* magazine, a gold award from the UK-based educational charity D&AD, and Object of the Year by *The Face* magazine. The team is responsible for the design of all Apple's products. 122–5

Ron Arad was born in Tel Aviv, Israel, in 1951 and studied at the Jerusalem Academy of Art and at the Architectural Association in London. In 1981 he co-founded One Off. In 1988 he won the Tel Aviv Opera Foyer Interior Competition and formed Ron Arad Associates the following year in order to realize the project. Other projects include furniture design for Poltronova, Vitra, Moroso, Alessi, The Gallery Mourmans and Driade and the interiors of the Belgo restaurants in London. In 2000 he had a major retrospective at the Victoria & Albert Museum, London. His work is included in several public collections worldwide and has been featured in numerous exhibitions and publications. Arad is currently professor of furniture design at the Royal Academy of Art in London. 22, 30

Antoni Arola was born in Tarragona, Spain, in 1961 and is one of the country's best-known designers. His work includes everything from a perfume bottle and a lamp to interior design projects. Arola was awarded with the Spanish National Design Prize in 2003. 166

Artek was founded in 1935 by four young Finns led by the visionary modern architect Alvar Aalto to market his innovative furniture, lamps and textiles internationally. It began as an industrial arts centre where the art and design trends of the time converged, and today the company, now led by Hendrik Tjaerby, is renowned as one of the most important contributors to modern design. Artek's most important products continue to be Aalto's furniture, lamps and textiles, and the company continually renews and reissues these, occasionally in the form of 'a challenge to the master'. In recent years, the company has also extended its collection by taking on new designers, and offers more than fifty designs. Artek has its headquarters in Finland, and is owned by the international investment group Proventus, based in Sweden. It has three showrooms in Finland (Helsinki, Espoo and Turku) and one in Stockholm. 37

François Azambourg has won much recognition over the years, including the Grand Prix du Design de Paris (2004), Lauréat du Concours Top Plastique (three times), and Lauréat du Concours du Musée des Arts Décoratifs. Currently, he is working on a sculptural chandelier for Galeries Lafayette, Toulouse, and with Japanese companies on the recycling of plastics. 29

Maarten Baas was born in Arnsberg, The Netherlands, in 1978 and studied at the Design Academy Eindhoven. His clients include Pol's Potten and Moooi. His designs have been exhibited and published many times. 36, 58

Barber Osgerby was formed in 1996 by Edward Barber and Jay Osgerby, who met while studying at the Royal College of Art in London. Since then, they have been developing design collections for leading manufacturers and clients, including Authentics, Magis, Isokon Plus and Stella McCartney. Their designs have been exhibited worldwide and have won numerous design awards, including the prestigious Jerwood Prize for the Applied Arts in September 2004. Their Loop Table is in the permanent collections of the Victoria & Albert Museum, London, and the Metropolitan Museum of Art, New York. In 2001, Edward Barber and Jay Osgerby formed Universal Design Studio, a multidisciplinary company specializing in architectural, industrial design and interior projects. 41, 42, 99

Massimo Barbierato was born in 1972 in Asiago, Italy, and received his degree in architecture and design from the University of Architecture in Venice. He works in the fields of architecture, design and visual communication and has collaborated with many architecture studios, art foundations and museums. Currently, he is an independent industrial product designer based in Venice, where he runs his own studio. He also works on architectural projects as well as interior design in Italy, Poland and the Czech Republic. Barbierato has won many international competitions and prizes. 84

Max Barenbrug, an industrial designer, formed Bugaboo in 1999 with his brother-in-law Eduard Zanen, a medical doctor and sales director. Max designed the first Bugaboo Frog for his final project at the Design Academy Eindhoven in 1994. He graduated with honours and was awarded Best Design of the Year. Bugaboo is a fast-growing company and its pushchairs are now sold in nearly thirty countries worldwide. 111

Tom Barker is the inventor of SmartSlab and CEO of SmartSlab Ltd and has worked in technology and design for the past sixteen years. He has extensive applied experience of projects utilizing advanced structures and new materials, as well as comprising electronics and digital software. Barker was a project design engineer for the capsules and boarding system of the London Eye Ferris wheel and developed photovoltaic solar products for BP Solar and new technologies for five zones in London's Millennium Dome. He has patented more than twenty inventions, a number of which have achieved commercial success, and is the youngest professor at the Royal College of Art in London, in the innovative department of Industrial Design Engineering. He also teaches an MA course at the Architectural Association's Design Research Laboratory, and his book on advanced design will be published in 2007. 162

Eric Barrett *See* Concrete Blond

Yves Béhar is a graduate of the Art Center College of Design in Pasadena, California, and was Design Leader at Frog Design and at Lunar Design before founding the San Francisco-based design studio, Fuseproject. In 2004 the San Francisco Museum of Modern Art exhibited a solo show of Béhar's diverse work, a 'futurespective' that spanned products, fashion, graphics, packaging, environments and strategy. Fuseproject's recent clients include Herman Miller, Birkenstock, Swarovski, Mini, Hussein Chalayan, Toshiba, Nike and Hewlett-Packard. 38, 120

Mario Bellini was born in 1935 and graduated from the Polytechnic of Milan, Italy, in 1959. His work ranges from architecture and urban design to furniture and industrial design. He became well known in the design community in 1963, when he worked as a consultant for Olivetti and Cassina. He has since won several Compasso d'Oro awards. Since 1980 Bellini has worked as an architect in Europe, the USA, Japan, Australia and the United Arab Emirates. From 1986 to 1991, he was the editor of *Domus*, the art, architecture and design review. Many of his designs are in the permanent collections of major museums, including twenty-five in the Museum of Modern Art in New York, and his work has been exhibited in numerous solo shows worldwide, most recently in 2005 at the Milan Triennale. Bellini has taught and lectured at numerous universities and international cultural institutions. 28

Noa Bembibre was born in A Coruña, Spain, in 1981 and studied at the School of Fine Arts in Bilbao before moving to Helsinki, Finland, in 2003 to study at the University of Art and Design. She established Noa Bembibre Tmi, a design and production studio based in Helsinki, in 2006. 178

Mathias Bengtsson studied at Art Centre College, Montreux, Switzerland, before taking a BA in furniture design at the Danish Design School, Copenhagen, and an MA in furniture/industrial design at the Royal College of Art in London. His work has been exhibited at prestigious shows and galleries throughout Europe, the USA, Canada and Brazil. 26, 35

Dror Benshenrit studied art and design at the Design Academy Eindhoven in The Netherlands, and the Centre for Art Education at the Tel Aviv Museum of Art in Israel. The Dror Studio has undertaken projects for Yigal Azrouel, Kiehl's and Material ConneXion (all in New York), and Edelkoort Studio and Trend Union (both in France). Benshenrit is dedicated to employing innovations arising from the use of new materials, techniques and shapes. 176

Pieke Bergmans, industrial designer, was born in The Netherlands in 1978. She graduated from the Design Academy Eindhoven in 1998 and received an MA in product design from the Royal College of Art in London 2004. From 2000 to 2005 she worked in design studios in The Netherlands, Germany, Italy and China. Bergmans's work has been exhibited internationally, and she had a solo show entitled 'You are Infected' at the Milan Furniture Fair in 2006. 152, 185, 212

Sebasatian Bergne (born 1966) studied industrial design at the Central School of Art and Design in London (1985–8) and then at the Royal College of Art, graduating with distinction in 1990. He now lives and works in London and Bologna, Italy. Since establishing his own studio in 1990, he has collaborated with many internationally renowned manufacturers of consumer products and furniture. His work has been widely published, exhibited, included in permanent collections at the Museum of Modern Art (New York) and the Design Museum (London), and honoured with numerous international design awards. Bergne contributes to design education in the UK, Switzerland and Italy. 186

Fabrizio Bertero graduated in architecture from the Polytechnic of Milan, Italy, in 1993. **Andrea Panto** and **Simona Marzoli** both graduated from the European Institute of Design in Milan in 1997, Panto in industrial design and Marzoli in interior architecture. The three now collaborate on architectural and industrial-design projects. They have undertaken architectural projects in Paris, Milan, Sydney, Seoul and Bangkok, and their furniture designs have been exhibited at the Milan Furniture Fair. Their Globule pouf for Zanotta is in the permanent collection of the Vitra Design Museum, Weil am Rhein, Germany. 49

Enzo Berti studied at the Accademia delle Belle Arti in Venice, under the guidance of Alberto Viani, and later attended the industrial design advanced course of the Architecture University in Venice. He designs interiors, furniture and a wide range of products, working with wood, metal and even sand. Among his clients are Artemide, Cinova, Gervasoni and Magis. During his career, he has received numerous awards. 24

Fabio Biancaniello completed his studies at ECAL, Lausanne, Switzerland, in September 2005 and now works in Lausanne. The Boschetto wardrobe he designed for the Nanoo collection by Faser-Plast earned him a nomination for the Switzerland Design Award. 35

Hrafnkell Birgisson *See* Sebastian Summa

Andreas Bittis was born in Essen, Germany, in 1967 and studied architecture at Rheinisch Westfälisch Technische Hochschule, Aachen, Germany. After graduating in 2001 he founded his own design consultancy in Aachen. Among his clients are Bernd Kniess Architekten und Stadtplaner, and Kalhöfer Korschildgen Architekten, both in Cologne, Germany, and Industriedesign NOA, Aachen–Milano, Germany–Italy. Bittis has lectured in Europe. 212

Kate Blee graduated from the Edinburgh College of Art in 1984 and established her own studio in 1986. She has been involved with Christopher Farr Handmade Rugs since the company was established in 1988. Her work includes many private and corporate commissions, including two altar cloths for St Giles Cathedral,

Edinburgh. In addition to designing rugs for Christopher Farr, she produces a range of hand-painted blankets, hand-dyed silk scarves and ties for Britain's Contemporary Applied Arts and the Crafts Council. 225

Marc Boase was born in 1973 and graduated from Brighton University, England, in design for production. He established a multidisciplinary studio in 1998, initially designing an award-winning range of ceramic products, and received the Smeg Design Award (1999), and the Arts Council Award (2001 and 2003), among others. His studio is currently engaged in the design of ceramics for the catering industry, including Gruppo, Itsu, L'etranger and Tiger Lil, and also creates furniture and interior projects for both the commercial and domestic markets. Boase's work has been exhibited internationally. 109, 175

Jörg Boner lives and works in Zurich, Switzerland, and studied product and interior design at the Basel School, graduating in 1996. He opened his own studio in 1999 and has developed products, furniture and lighting for clients such as the German firms ClassiCon and Nils Holger Moormann, the Dutch company Hidden, and the Swiss company Nanoo by Faser-Plast. He has received a number of international design awards, including the iF (International Forum Design) gold award. Since 2001 Boner has also taught industrial design at the Lausanne Academy of Applied Arts. 145

Tord Boontje was born in 1968 in Enschede, The Netherlands. He studied industrial design at the Design Academy Eindhoven before moving to the UK to study at the Royal College of Art in London. He uses industrial technologies to create exquisite glassware, lighting and furniture. 71, 153, 180, 188

Fabio Bortolani was born in Modena, Italy, and graduated in architecture in Florence. Clients have included Agape, Alessi, Authentics, Driade, La Palma, Montina, Pandora Design and Elmes. Bortolani has won a number of awards, including the Promosedia Top Ten prize, Udine (1992 and 2003) and the Design Plus award, Frankfurt (1997), and his work was selected for the Compasso d'Oro, Milan (2001). 178

Ronan and Erwan Bouroullec are brothers who have collaborated since 1998. Erwan studied industrial design in Paris at the Ecole Nationale Supérieure des Arts Appliqués and the Ecole Nationale Supérieure des Arts Decoratifs. Ronan graduated in applied and decorative art and has worked freelance since 1995, designing objects and furniture for Liaigre, Cappellini, Ligne Roset and Galerie Neotu. He was awarded Best Designer at the ICFF, New York in 1999. 34, 40, 58, 59, 64, 78, 168, 208

Brand van Egmond, based in The Netherlands, is a design partnership between William Brand and Annet van Egmond. William Brand (b. 1963) and Annet van Egmond (b. 1964) graduated from the Hogeschool voor de Kunsten in Utrecht, The Netherlands, where he trained as an architect and she as a sculptor. Their lighting, furniture and accessories have been shown at fairs and galleries in Cologne, Toronto, Milan, Frankfurt, Rotterdam, London, Paris, Kortijk (Belgium) and Tokyo and can be found in prestigious museums, embassies, restaurants, hotels, parliament buildings and theatres around the world as well as in the homes of celebrities. 157

Andrea Branzi is an Italian architect, designer and critic, who was born in Florence but now lives and works in Milan. He was a member of Archizoom Associati, the Italian avant-garde group of radical design, until 1974, and in 1983 was among the founders of the Domus Academy, the postgraduate school of design in Milan. In 1987 he was awarded the Special Compasso d'Oro for his work as a designer and theorist. He has held many exhibitions of his work, including at the Musée des Arts Décoratifs in Paris and at the Museum of Modern Art in New York. He is a professor in the Third Faculty of Architecture and Industrial Design of The Polytechnic University of Milan, lectures internationally and has written several books on design. 83, 182–3

Cam Bresinger is a graduate of Rhode Island School of Design in Providence. He has designed space suits and is the president of Nemo Equipment. **Suzanne Turell** is a graduate of Princeton University and the Rhode Island School of Design, who began working for Nemo shortly after graduating with an MA in interior design in 2005. Nemo has won numerous design awards, including the iF (International Design Forum) award, Hannover, Germany, and recognition from *Popular Science Magazine* for producing one of the best inventions of 2005. 92

Maria Luisa Brighenti is an Italian designer who, since the 1970s, has specialized in interior design and product research and communication consultation. Her primary research interest has always been ceramics, and she has continually sought to develop products to be used in new ways in interior architecture. I Mosaici, which she designed for Ceramica Vogue, received the Design Auswahl award from the Stuttgart Design Centre, Germany, in 1989. 204

Persijn Broersen and Margit Lukács are based in Amsterdam, The Netherlands. They studied together at the Rietveld Academy and at the Sandberg Institute in Amsterdam, where they each received an MA in free arts and design. Their work covers a broad range of arts, from video installations, drawings, murals and films to commercials and music videos. 214

Debbie Jane Buchan was born in Edinburgh, Scotland, in 1973. She studied textiles at Gray's School of Art at Robert Gordon University in Aberdeen, graduating with an MA in surface communication in 1996. Her work is concerned with 'smart' fabrics and pushing the boundaries between design disciplines through the innovative use of applications and technologies. 200–1

Stephen Burks was born in 1969 in Chicago, Illinois, and studied product design at the Illinois Institute of Technology's Institute of Design and architecture at Columbia University's Graduate School of Architecture and Planning. He formed Readymade Projects in 1997 to offer creative design direction and design strategy in a variety of disciplines. Since then, he has collaborated with some of the most interesting brands, designers and artists working today, for clients as diverse as Idee's Sputnik and Cappellini. 156

Büro Für Form was founded by Benjamin Hopf, Constantin Wortmann and Alexander Aczél in Munich 1998. The group specializes in interiors, furniture, lighting, tableware and luxury accessories and its design philosophy is to combine organic shapes and geometric elements with humour to create a particular modern style. Clients include Siemens, Osram, Habitat, Koziol, Next and Red Bull. The company has won a number of awards, including the iF (International Design Forum) award, Hannover, Germany, 2005. 142

Fernando and Humberto Campana have been working together, based in São Paolo, Brazil, since 1983. The brothers, who produce political rather than merely functional objects, were thrust into the limelight with their 1989 exhibition 'The Inconsolable'. Their radical vision involves using traditional materials and industrial surplus to create an independent Brazilian design language. 20, 46

Louise Campbell was born in 1970 and graduated with a degree in furniture manufacture and innovation from the London College of Furniture in 1992, then from the Institute of Industrial Design in Denmark in 1995. She has developed several products for the Bahnsen Collection, Denmark, and for Louis Poulsen Lighting and has also been commissioned by the Danish Ministry of Culture, Hansen & Sørensen and Kolding Kommune, among others. Her work has been exhibited internationally and has received numerous awards, including the iF (International Forum Design) gold award in 2005 for Campbell Pendant. 25, 70, 146

Giulio Cappellini was born in 1954 and received a degree in architecture from Bocconi University in Milan, Italy. In 1979 he joined his family's firm, Cappellini, and as art director transformed it into a springboard for important international designers, as well as designing furniture and objects himself. His work includes a collaboration with Ceramica Flaminia to design new products. 133

Marek Cecula was born in Poland in 1944 and lives and works in New York and Poland. He was the head and coordinator of the ceramic department of the Parsons School of Design in New York from 1985 to 2004, and has been a professor at the National College of Art and Design in Bergen, Norway, since 2004. His work has appeared in many publications and he has received numerous fellowship grants and awards. His designs have been exhibited in numerous group and solo shows worldwide, and are in the permanent collections of various museums, including the Smithsonian National Museum of Art in Washington DC and the Museum of Modern Art in New York. 175

Antonio Citterio was born in Meda, Italy, in 1950 and received a degree in architecture from the Polytechnic University of Milan. He opened his studio in 1972, mainly focussing on industrial design, and now has clients such as the Albatros–Sanitec Group, Arclinea, Axor–Hansgrohe, B&B Italia, Flexform, Flos, Fusital, Guzzini, Iittala, Kartell, Technogym and Vitra. In 1999, Citterio formed a partnership with Patricia Viel, a multidisciplinary studio for architectural design, industrial design and graphics. The studio develops projects for residential complexes and trade centres, industrial sites, the restructuring of public buildings, and the planning of workspace, offices, showrooms and hotels. It is also operates in the field of corporate communications and implements corporate image projects. Citterio has won numerous awards, including the iF (International Design Forum) award, Hannover, Germany, for the tableware collection My Table designed for Guzzini. 33, 54, 63, 114

Cesare (Joe) Colombo was born in Milan, Italy, in 1930. He studied painting and sculpture at the Academy of Fine Arts in Brera, Milan, before studying architecture at the Polytechnic University of Milan from 1951 to 1955. From 1962, after working as a painter and sculptor, he dedicated himself to the design of products for industrial mass production. Colombo opened a design studio in Milan and worked on architectural and design commissions for clients such as Alessi, Boffi, Kartell, Stilnovo, Rosenthal and Zanotta. His work won many awards and is in the permanent collections of numerous museums worldwide, including the Museum of Modern Art in New York. Colombo died in 1971, at the age of forty-one. 101

Committee Design Studio was formed by **Clare Page** and **Harry Richardson**, both born in 1975, who met while studying fine art at Liverpool Art School, England. After graduating, Page studied machine knitting and Richardson hand-crafted cabinetmaking. They married and formed the Committee Design Studio in 2001, after taking on a derelict shop in Deptford, southeast London, which they developed into the Gallop studio and gallery space. Their company designs furniture, lighting, textiles, wallpaper and interiors, and their clients include Topshop, Liberty's and the Crafts Council. Committee's work has been exhibited at the Design Museum, London, and presented at the Milan Furniture Fair, among other venues. 212–3

Concrete Blond specializes in interior and exterior wall cladding, flooring and bespoke architectural concrete castings. The company was founded in 2000 in London by Eric Barrett, who had been living the life of a 'dumpster-raider' in Newcastle in the north of England, close to where he had studied glass, ceramics and 3D design in Sunderland. Originally his focus was on finding new uses for industrial detritus and experimenting with the possibilities of old forms with new functions, but after creating a comprehensive portfolio of art and design pieces, he moved to London and set up Concrete Blond as a reaction to the prevailing conservatism in the use of concrete in interior and exterior spaces. Since exhibiting at Designersblock 2001, Concrete Blond has established a strong track record in producing beautiful, witty and unique interior spaces. 207

Carlo Contin was born in Milan, Italy, in 1967 and, after attaining his degree in 1987, specialized in interior architecture at the Advanced Institute of Architecture and Design in Milan. He opened his own studio in 1998. His work has been exhibited at the Salone Satellite and other venues in Milan since 1999, and at the Biennale Internationale Design 2002, Saint-Étienne, and other European shows. His clients include Cappellini, Coop, Meritalia, Slamp, and the Metropolitan Museum of Art in New York. 95

Christopher Coombes was born in Bristol, England, in 1979. In 2001 he graduated from Brunel University, London, with a degree in industrial design. The same year he moved to Milan, Italy, and began collaborating with George Sowden, then went to work with Sebastian Bergne, during which time he was also developing his own projects. For the past two years he has developed projects with Cristiana Giopato from their studio in Milan. 86

Aki and Arnault Cooren are a Japanese–French designer couple based in Paris. Their work encompasses retail shop design, scenography, product design and furniture, and includes the flagship store of Shiseido in Paris, perfume bottles for L'Oréal Luxury Products and lights for Artemide. 46

Dominic Crinson trained as a fine artist and practised ceramic art for nineteen years, producing large, sculptural hand-glazed vessels and exhibiting his work internationally. He then began hand-making and -firing decorative ceramic tiles, which led to the development of his technique of processing digital images on ceramic tiles. Crinson has a broad international customer base and has worked for clients from Australia to Russia. In 2004 he was asked to produce a design for the 'Bombay Sapphire 'Inspired' advertising campaign, and he was named Best Designer in the Designer's Choice Zone of the Seoul Living Design Fair 2006. 223

Lorenzo Damiani was born in 1972 and graduated in architecture from the Polytechnic of Milan, Italy, and, with a master's degree, from the Polytechnic School of Design, Milan. His work has been shown in collective as well as two solo exhibitions both in Italy and abroad. Damiani has won several awards and competitions and has collaborated with various companies, including Campeggi, Cappellini, Firme di Vetro, Acqua di Parma, Abet Laminati, Omnidecor, Coop. 83, 96, 138

Carlotta de Bevilacqua graduated in architecture from the Polytechnic University of Milan, Italy, in 1983. From 1989 to 1993 she was the art director of Memphis and Alias, and performed research on artificial illumination, producing innovative products for Artemide. In 1999 she purchased Danese, the leader in Italian design in the 1960s and '70s, and revived the company. Until 2004 she was chief executive officer for the brand strategy and development organization within the Artemide Group, involved in marketing, communication and vision, and strategic futures. From 1999 to 2001, De Bevilacqua was a university lecturer in industrial design at the Polytechnic University of Milan, and from 2001 to 2004 she was a lecturer in the design MA programme at Domus Academy in Milan. 154

Judith de Graauw *See* Studio Demakersvan

Christophe de la Fontaine was born in Luxembourg in 1976 and studied sculpture at the Luxembourg Lycée des Arts et Métiers (1993–6) and industrial design at the Stuttgart State Academy of Art and Design, Germany, (1996–2002) before opening a design office in Munich with Stefan Diez. He now lives and works in Milan, Italy, where he opened his own studio in 2003. 43

Barbara de Vries was born in The Netherlands in 1982. She studied at the Design Academy Eindhoven, holding internships at Moooi and Marcel Wanders Studios, and graduated in 2005. Following her graduation, she began work on prototypes for a series of decorative electrical sockets, as well as a children's clothing collection and associated window and shop displays for Vroom & Dreesmann in The Netherlands. 117

Tim Derhaag was born in 1974 in Geleen, The Netherlands, and graduated in 1997 in product design from the University of Arts, Arnhem. From 1997 to 1999 he attended the Royal College of Art in London and received an MA in furniture and industrial design in 1999. In 2000, Derhaag established his own design studio in London. His work includes the design of the exhibition stand and retail shop concept for the Buddhahood fashion label and for 10 to 10 by Dexter Wong (2001), the design of a collection of office furniture for Ontwerpzaken in Utrecht (2002), and the design of the 'shop-in-shop' concept and the Zurich showroom for Modular Lighting Instruments lighting company (2002 and 2003, respectively). 161

Christian Deuber was born in Switzerland in 1965, received his degree in engineering in 1993 and lives and works in Lucerne. He co-founded N2 Design, and produces designs for lighting and products. He has developed furniture and lighting for design companies such as Palluccoitalia and Driade in Italy, Hidden in The Netherlands and Nanoo in Switzerland, among others and has won numerous awards, including the iF (International Design Forum) award, Hannover, in 2001, 2003 and 2005. 145

Stefan Diez was born in Freising, Germany, in 1971. After a course in architecture and an apprenticeship as a cabinetmaker he went to India for a year where he designed and constructed furniture for a company in Bombay and Poona. Returning to Germany, he studied industrial design at the Academy of Fine Arts in Stuttgart (1996–2002). He then worked as assistant to Richard Sapper and later for Konstantin Grcic. In 2003 he set up his own design studio in Munich where he develops furniture, products and exhibition designs. Clients have included Rosenthal, Authentics, Elmar Flötotto, Nymphenburg Porzellanmanufaktur, Schönbuch, WMF, Uwe Braun/Würth Solar, Fischer and Nanoo. 24, 42, 43, 178–9

Jan Jannes Dijkstra was born in 1982 in Olterterp, The Netherlands. He graduated from the Design Academy Eindhoven in 2005 and began working as a freelance designer for Prins Bernard Cultuurfonds, Amsterdam, and Almat kitchens Oisterwijk. His work has appeared in several exhibitions and publications. 101

Tom Dixon was born in Sfax, Tunisia, in 1959 and moved to the UK when he was four years old. A self-taught designer, he has been creative director of Habitat UK since 1997, and also runs his own design studio, Tom Dixon, which he founded with David Begg in 2002. Since its inception, the studio has developed its own collection of contemporary lighting and furniture, including the acclaimed Mirror Ball Collection of lights. In 2004, a partnership was established between the company and private investment firm Proventus, forming Design Research, which today owns and manages both the Tom Dixon brand and Artek, the Finnish modernist furniture manufacturer established by Alvar Aalto in 1935. Dixon was awarded an OBE for services to British Design in 2000. 25, 141, 146, 151, 164

Sylvia Döhler was born in 1972. Before beginning to study textile design, she worked as a florist. Since graduating, she has created Organic Non-wovens, textiles that function as a bridge between her work as a florist and as a designer. 215

Doshi Levien is a young London-based partnership led by Jonathan Levien and Nipa Doshi, who met in London at the Royal College of Art, graduating in 1997. After working for prestigious design offices in London, Milan and India, they married and set up Doshi Levien in 2000, immediately receiving a commission from Tom Dixon, creative director of Habitat. 102

Maurizio Duranti was born in Florence, Italy, in 1949 and graduated in architecture in 1976. Between 1984 and 1989 he tutored at the Faculty of Architecture in Genoa and at the European Institute of Design in Milan. His work has included architecture, interior design and industrial design. Since 1990 he has focussed on the latter and has developed many successful household products for leading manufacturers. His work has been exhibited internationally and is in the permanent collections of many museums worldwide, including the Athenaeum in Chicago, the Bunkamura Design Collection in Tokyo, and the Victoria & Albert Museum in London. Among his clients are IB Rubinetterie, Villeroy & Boch, Ancap, Barazzoni, Gallotti & Radice and Vitruvit. In addition to numerous articles about his work published worldwide, he has published two monographs. 134

El Ultimo Grito was founded in 1997 by Roberto Feo and Rosario Hurtado. Roberto Feo was born in London in 1964 but grew up in Madrid, Spain. He studied furniture design at the London College of Furniture and completed an MA in furniture Design at the Royal College of Art. Rosario Hurtado was born in Madrid in 1966 and moved to London in 1989 to study cabinet-making and furniture design at the London College of Furniture. She completed her BA in industrial design at Kingston University, London. 21, 40

Susan Elo, a graduate of the Helsinki University of Art and Design, works in lighting, furniture and interior design and has had her own studio, Muotoilu Elo, in Helsinki since 1998. She has cooperated with Sirpa Fourastié on the Kuutio futon, which was presented at the Milan Furniture Fair 2005. 88

Alexander Estadieu is from France and studied product design for several years there before moving to Helsinki, Finland, to complete an MA in applied art and design at the Helsinki University of Art and Design. He is now based in Helsinki as a product designer. 191

Mieke Everaet was born in Belgium in 1963. From 1981 to 1985 she studied in the ceramic department of the Royal Academy of Fine Arts in Antwerp. She graduated from the National Higher Institute of Fine Arts in Antwerp in 1988. Everaet has participated in numerous international exhibitions and is represented in several large museum collections. She has received many awards and prizes in Belgium and abroad. 194

Lotta Fagerholm is studying for a master´s degree in the Department of Applied Art and Design at the Helsinki University of Art and Design, Finland, and has a bachelor's degree in ceramics and glass. 99

Estefanía Fernández, who lives and works in Milan, Italy, was born in 1978 in Spain. After she graduated with a degree in interior design in Madrid, she worked in several interior design firms in Spain. In 2005 she moved to Milan to take the industrial design MA course at the Polytechnic School of Design. Fernández plans to continue in the field of interior design and scenography, and to also design furniture. 69

Marta Daza Fernández was born in 1975 in Madrid, Spain, and discovered her gift for design at the Accademia d'arte Maite Cuaresma. In the late 1990s she collaborated on scenography and interior design projects in Spain, then in 2000 she moved to Italy, where she studied at the prestigious Polytechnic School of Design in Milan. After graduating, she settled in Milan and in 2005 began her collaboration with Serralunga. In the field of textiles she works together with industrial designer Ilaria Marelli. Concurrently, she contributes to Ansbacher studio in Milan on interior design projects. 185

Christian Flindt is a Danish designer whose work is about communication and social interaction, conditions between people and objects in a spatial environment. His work has been exhibited in Europe and Japan and has won several awards, including the Design Prize 2005 at the Copenhagen International Furniture Fair. He was voted Designer of the Year 2006 by *Bo Bedre*, Denmark's largest lifestyle magazine. 28

Gwendolyn Floyd *See* Ransmeier and Floyd

Johannes Foersom worked as a cabinet-maker at Gustav Berthelsen, Copenhagen, Denmark, before attending the Arts and Crafts School in Copenhagen in 1972. He began a partnership with Peter Hiort-Lorenzen in 1977. His work, independently and with Hiort-Lorenzen, has received several awards. 62

Jacopo Foggini was born in Turin, Italy, and currently lives and works in Milan. While working in his family's plastics business he discovered the versatility of methacrylite, an industrial material normally used to make automobile tail-lights. Fascinated by the extraordinary aesthetic and chromatic qualities of this material, he began to experiment with it at the beginning of the 1990s. Foggini's work has been exhibited in more than thirty galleries and venues worldwide, including the Carrousel du Louvre and the Centre Georges Pompidou in Paris. His lighting sculptures are in the permanent collections of the Haus der Musik in Vienna, Austria, and the Montreal Museum of Decorative Arts, Canada. He collaborates with some of the world's most renowned architects, and designs projects for hotels, museums, showrooms, private residences and public spaces. He is currently focussing on several projects in the Middle East, Japan and the USA. 163

Jozeph Forakis was born in New York in 1962, an American of Greek and Russian origin. He graduated from the Rhode Island School of Design, then studied at the Domus Academy in Milan, Italy, where he has now lived for the past ten years. Although he has a penchant for things technological, he also designs furniture and lighting. 161

Formstelle studio for architecture and design was founded in 2000 by Claudia Wiedemann and Jörg Kürschner. The studio works in the fields of architecture, product design and corporate identity, and its work has been exhibited at the international furniture fairs in Milan and Cologne. 49

Foster & Partners, world-famous architectural practice, was formed by Norman Foster in 1967. The partnership has received more than 300 awards and citations for excellence and has won more than sixty national and international competitions. Lord Foster, who is a founder trustee of the Architecture Foundation of London, taught architecture in the UK and the USA, and has lectured all over the world. The work of his practice has been exhibited internationally, and forms part of the permanent collections of the Victoria & Albert Museum in London, the Museum of Modern Art in New York, and the Centre Georges Pompidou in Paris. 34, 105

Nick Foster graduated from Brunel University, London, in 1998, and began work as a design engineer developing products for Dyson Research. He gained his MA in product design in 2001, and joined Sony as a senior designer in 2005. He has worked with a variety of international clients and has been involved in the design, research and development of many household products, from mobile phones to washing machines. 113

Sirpa Fourastié was born in Kuopio, Finland, in 1970. She studied in the Faculty of Art History of the University of Helsinki (1990–1) and at Les Ateliers des Beaux-Arts in Paris (1994–7) and received an MA in art education from the Helsinki University of Art and Design, where she is currently studying for a PhD in design. In addition to teaching, she is manager, producer and designer of the Kuutio futon, a design cooperation with Susan Elo, which was presented at the Milan Furniture Fair 2005. Her work has been exhibited in Finland, North America and Italy. 88

Enrico Franzolini was born in Udine, Italy, in 1952 and studied in Florence and at the University of Venice. He had his first solo exhibition at the Plurima Gallery in Udine in 1978, which marked the beginning of his interest in minimalist architecture. During the same period, he also became interested in interior and product design, and in the 1980s and '90s some of his architectural and industrial design projects were published in specialist magazines such as *Domus*, *Modo*, *Abitare* and *Interni*. In 1993 his project for a private house won the Piranesi Prize. He has worked for clients such as Accademia, Alias, Cappellini, Crassevig, Knoll International, Montina and Pallucco. 146

Front is a design group based in Stockholm, Sweden. The members are Sofia Lagerkvist, Charlotte von der Lancken, Anna Lindgren and Katja Sävström. Front works conceptually with design and often lets external factors affect the process of design. The firm's work has been shown worldwide and it has had solo exhibitions in Stockholm, Amsterdam, London and Milan. *Wallpaper* magazine gave the group the Best Young Designer award in 2005. 58, 72, 120, 184

Naoto Fukasawa was born in 1956 in Yamanashi, Japan, and graduated from Tama Art University, Tokyo, in 1980 with a degree in product design. He worked for IDEO America, then set up the IDEO Tokyo office in 1996 before establishing his own company in 2003. He now teaches industrial design at Tama Art University and is director of Tokyo AAD Studios. He has received more than forty European and American design awards and is a member of the Japanese Ministry of Economy, Trade and Industry's Strategic Design Research Society. 47, 52, 59, 62, 64, 66, 104, 121, 132

Massimiliano Fuksas, architect and designer, was born in Rome in 1944 where he graduated in architecture from La Sapienza University in 1969. He established practices in Rome, Paris and Vienna in 1967, 1989 and 1993, respectively, and has recently opened a new studio in Frankfurt. He has been visiting professor at universities in Germany, France, Austria and the USA and has received numerous awards for his work. 225

Masahiro Fukuyama was born in Kumamoto, Japan, in 1976 and studied at Tokyo National University of Art and Music, graduating in 2002, and is pursuing the Sandberg Institute fine art master course. 145

Mark Gabbertas, an award-winning furniture designer based in London, worked in advertising before making a career transition and training to be a cabinet-maker more than fifteen years ago. He began as a designer-maker, undertaking both public and private commissions. Recently, he has concentrated on the design of products for various European and international manufacturers. His clients include Allermuir, Lloyd Loom and Boss Design, and he has been commissioned to design and manufacture furniture for the Tate Gallery, M & C Saatchi and the Virgin Group. Gabbertas' work has been shown in numerous exhibitions worldwide. 211

Aleksandra Gaca was born in 1969 and graduated from The Hague Royal Academy of Art, The Netherlands, with a degree in textile and fashion design, in 1997. She received a number of grants from the Netherlands Foundation for Visual Arts, Design and Architecture (Fonds BKVB), which enabled her to continue her research on three-dimensional weave structures. Gaca's fabrics are in a number of museum collections and have received various international awards, such as the Decosit Awarda (Belgium), the Dutch Design Award and the Neocon Award (USA). 228

Jesús Gasca was born in 1939 in San Sebastián, Spain. In 1982 he founded the company STUA, where he has been developing his design work. 44

Frank Gehry, who was born in 1929 in Toronto, Canada, as Frank Owen Goldberg, is widely considered one of the finest and most artful of contemporary architects. His most important and acclaimed building to date is the Guggenheim Museum in Bilbao, Spain (1997), a large structure of voluptuous, swooping, organic forms covered in gleaming titanium steel that made him an international star. Gehry also designs furniture and other utilitarian objects. Prominent among his many awards are the Pritzker Prize (1989) and the first Gish Award (1994). 168

Michael Geldmacher *See* Eva Paster

Ron Gilad was born in Israel and now lives in New York. He graduated with a degree in industrial design from the Bezalel Academy of Art and Design, Jerusalem, in 1998. From 1999 to 2001, Gilad was a lecturer in 3D and conceptual design in the jewellery design department at Shenkar Design and Engineering Academy in Ramat-Gan, Israel. In 2001 he co-founded Designfenzider, LLC, in New York. He has guest-lectured at the School of Visual Arts in New York and other US art and design schools. His work, which ranges from furniture and jewellery to entire interiors, has won many awards, has been widely exhibited in both solo and group exhibitions since 1999 and is in the permanent collections of the Museum of Modern Art in New York, the Museum of Arts and Design in New York and the Tel-Aviv Museum of Modern Art in Israel. 147

Cristiana Giopato was born in Treviso, Italy, in 1978, and moved to Milan to continue her studies at the Polytechnic of Milan, graduating in 2002. During her studies she worked for Studio Hasuike, and since 2003 has collaborated with Patricia Urquiola. For the past two years she has developed projects with Christopher Coombes from their studio in Milan. 86

Stefano Giovannoni was born in 1954 in La Spezia, Italy. He graduated in architecture in Florence in 1978 and now lives and works in Milan. Since 1979 he has been teaching and doing research at the University of Florence Faculty of Architecture and has been Master-Professor at the Domus Academy in Milan and at the Università del Progetto in Reggio nell'Emilia. He also works as an industrial and interior designer and architect, specializing in plastic products. Giovannoni has won a number of awards and his works are part of the permanent archive of the Centre Georges Pompidou in Paris and of the Museum of Modern Art in New York. 51

Konstantin Grcic was born in Munich, Germany, in 1965. After training as a cabinetmaker at Parnham College, Dorset, England, he studied design at the Royal College of Art in London. Since setting up his own design practice, Konstantin Grcic Industrial Design, in Munich in 1991, he has developed furniture, products and lighting for some of Europe's leading design companies, including Authentics, Cappellini, Driade, Flos, Iittala, SCP and Whirlpool. Many of his products have received prestigious design awards: his Mayday-lamp, produced by Flos, was selected for inclusion in the permanent collection of the Museum of Modern Art in New York and won the Compasso d'Oro in 2001. 39, 42, 54, 98, 104, 119

Gudrun Lilja Gunnlaugsdóttir studied furniture making in Reykjavik, Iceland (1991–4), and later at the Akureyri School of Art, Iceland (2000–2) and at the Design Academy Eindhoven, The Netherlands (2002 –5). She has worked and taught both in The Netherlands (including with Jurgen Bey) and Iceland and, since 2005, has run her own design practice, Studio Bility, based in Iceland. Her work has been exhibited throughout Europe (Italy, Spain, Germany, Iceland, The Netherlands) and has been published in several magazines, journals and books. 68

Alfredo Häberli was born in 1964 in Buenos Aires, Argentina. He moved to Switzerland in 1977, where in 1991 he graduated in industrial design from the Höhere Schule für Gestaltung in Zurich. He received the Diploma Prize, SfGZ, in 1991. From 1988 onwards he worked in Zurich for the Museum für Gestaltung, where he was responsible for numerous exhibitions. In 1993 he set up his own studio, and subsequently worked for firms such as Alias, Authentics, Edra, Driade, Luceplan, Thonet and Zanotta. Recently Häberli has developed products for Asplund, Bd Ediciones de Diseño, Cappellini, Classicon, Iittala, Leitner, Moroso, Offecct and Rörstrand. Häberli's designs have been shown in numerous exhibitions throughout Europe and he has received many awards for his work. 56, 80, 196, 225

Zaha Hadid, born in Iraq in 1950, is a world-renowned architect whose work was awarded the highest architectural accolade, the Pritzker Architecture Prize, in 2005. Her work is characterized by fragmented geometry and bold fluid forms, which she brings to her designs for furniture, lighting and kitchens, as well as her architecture. Among her numerous landmark projects is the building for BMW in Leipzig, and she is currently working on the National Centre of Contemporary Arts in Rome and an opera house for Guangzhou, China. Additionally, she has recently produced a conference/dining table and a chandelier for Established & Sons, London. Hadid is concerned with all aspects of architecture and design, including practice, theory, teaching and research. 103, 158–9

Knut and Marianne Hagberg, who are brother and sister (born in 1949 and 1954 respectively), both studied architecture in Copenhagen, Denmark. They have been working as designers since 1979, are members of the National Association of Swedish Interior Designers and have received the Excellent Swedish Design award for four of their products. 93

Ashley Hall was born in Cardiff, Wales, in 1967 and studied furniture design at Nottingham Trent University and the Royal College of Art in London, receiving his MA in 1992. He then worked as a designer in the fields of furniture, product, lighting and interiors for a variety of design consultancies and manufacturers. The furniture and product design consultancy he established in 1994 has now become part of Diplomat, the London-based partnership he runs with Matthew Kavanagh. Hall also works as design director for a bathroom-product manufacturer and as a visiting lecturer in product design at a number of British universities. 44–5, 94

Rosemary Hallgarten was born in 1966 in London, England, but moved to the USA in 1996 and currently lives in Westport, Connecticut. She is a second-generation craftsperson, following in the footsteps of her mother, Gloria Finn, who partnered with artists such as Gio Ponti and Anni Albers to interpret their paintings as floor coverings. From jewellery-maker in London to textile artist, Hallgarten has always celebrated the tactile sensuality of materials in her designs, and her new collection of rugs explores her ongoing creed of combining craft with a contemporary aesthetic. She works with artisans in Peru, Brazil and Nepal to create rugs, pillows and throws that resonate with a sense of place and craftsmanship. 209

Isabel Hamm acquired her MA in ceramics and glass from the Royal College of Art in London in 1998 and set up her own studio the same year. She produced her first designs for German company WMF in 1999, and undertook projects for Salviati Glass and Erler and Wieinkauff, and chandelier projects in a number of locations, including Berlin, Cologne and Madrid. 152, 167

Ineke Hans was born in The Netherlands in 1966. After graduating from the Hogeschool voor de Kunsten in Arnhem (NL) in 1991 and the Royal College of Art in London in 1995, Hans worked for three years for Habitat UK before returning to The Netherlands to found her own studio, Ineke Hans/Arnhem, in 1998, designing everything from jewellery to public-space projects. She won a Red Dot Award and Design Plus Award for a surprising garlic crusher for RoyalVKB in 2005 and another Red Dot Award in 2006 for a bowl and spoon, also for RoyalVKB. Her work has been purchased by international design collections and her clients include Swarovski, Cooper Hewitt Museum, New York, and MVRDV–architects. 173, 226

Marc Harrison was born in Takapuna, Auckland, New Zealand, in 1970. His design studio is based in Brisbane, Australia, where he studied interior design between 1989 and 1992 at the Queensland College of Art, Griffith University. Harrison has a strong interest in sustainable design and extensively researches and incorporates bio-composites into his products. His practice undertakes experimental design for international exhibitions, and also the creation of functional design, focussing on innovative lighting, homewares, architectural fittings and furniture for interiors and streetscapes. His commissions include door hardware for the Queensland University of Technology and street-furniture designs for the Gold Coast Convention Centre and Brisbane City Council. His work has received awards and has been shown in both group and solo exhibitions. 182

Stuart Haygarth was born in Whalley, Lancashire, England, in 1966 and studied graphic design and photography at Exeter College of Art and Design in Devon. Since 1991 he has worked as a freelance photographic illustrator, and since 2004 has been working on design projects that revolve around objects that are normally collected in large quantities, categorized and assembled in a way that transforms their meaning. The finished piece of work takes various forms, such as chandeliers, installations, and functional and sculptural objects. 169

Jaime Hayón was born in Madrid in 1974 and trained as an industrial designer in Madrid and Paris. In 1997 he began working as a researcher in Fabrica, Benetton Group's communication research centre in Treviso, Italy. A year later, he was appointed head of the design department, where he oversaw the development of interiors for shops, exhibitions and restaurants. In 2004, he began his career as an independent designer, with projects ranging from toys to furniture and interior design, as well as art installations. His furniture has been exhibited in the David Gill Gallery, London, and has also been shown at the Vitra Design Museum in Weil am Rhein in Germany and at the Design Museum, London. His clients include Benetton, Metalarte, Artquitect Edition, Coca-Cola, Danone Group, BD Ediciones, adidas, Palluco, Piper and Camper. Hayón's designs have appeared in all of the major design publications. 21, 136, 137, 180–1

Sam Hecht was born in London in 1969 and studied industrial design at the Royal College of Art. After training in the studios of David Chipperfield, he spent seven years working in the USA and Japan before being invited to become head of design at IDEO in London. In 2002 he co-founded Industrial Facility with his partner Kim Colin. His work has aroused great international interest, with his first solo exhibition presented in Paris in 2003. His clients include Whirlpool, Magis, Epson, Lexon and Muji Japan, and he has been both retained designer and advisor for the latter since 2001. His work has received more than thirty international awards, including two iF (International Forum Design) gold awards, and is part of the permanent collections of the Museum of Modern Art in New York and the Centre Georges Pompidou in Paris. Hecht lectures in several countries, including Japan, Germany, The Netherlands, Israel and the USA, and has written and contributed to many books. 94, 119, 127, 175, 178

Scott Henderson was born in Alexandria, Virginia, USA, in 1966. After graduating from Philadelphia College of Art and Design with a degree in industrial design in 1988, he worked for a variety of design consultancies, designing everything from furniture to lighting and even aircraft interiors. He joined Smart Design in 1993 where he is currently the director of industrial design. In addition to his many projects with Smart Design, Henderson works on solo projects ranging from furniture to ceramics. 109

Sebastian Hepting was born in Munich, Germany, in 1975 and graduated as an industrial designer from the Fraunhofer Institut, Freiburg, Germany, in 1999. He was an intern at Vitra, Basel, Switzerland, during the same year and from 2000 to 2002 collaborated with Vladimir Kagan in Italy and New York. Since 2003 he has been working for Ingo Maurer in Munich. 167

Herzog and de Meuron is a Basel-based Swiss architecture firm, founded in 1978 by Jacques Herzog and Pierre de Meuron (both born in 1950), its two main partners. In 2006, *New York Times* magazine called the firm 'one of the most admired architecture firms in the world'. The architects often cite Joseph Beuys as an enduring artistic inspiration and collaborate with different artists on each architectural project. Their success can be attributed to their skills in revealing unfamiliar or unknown relationships through familiar materials. Among their most notable projects are the Laban Dance Centre, Deptford Creek, London (2003), for which they won the RIBA Stirling Prize; Tate Modern, Bankside, London (1995–2000); and Dominus (winery), Napa Valley, California (1999). 168

Peter Hiort-Lorenzen worked as a ship carpenter in Elsinore, Denmark, before studying at the Royal Academy of Fine Arts in Copenhagen. In 1972 he established his own design studio, and five years later began a partnership with Johannes Foersom. His work, independently and with Foersom, has received several awards. 62

Jan Hoekstra was born in Maastricht, The Netherlands, in 1964. He trained as an industrial designer at the University of Genk, Belgium, and after graduating in 1989 worked at Flex/the Innovationlab in Delft for several years, where he designed and developed various successful consumer products. In 1993 he began teaching at the Design Academy Eindhoven in The Netherlands. Since 1997 he has worked for Royal van Kempen & Begeer and, as design manager, is responsible for product development of the popular Dutch cookware brands BK and Q-linair, and the international brand RoyalVKB. He is also currently design manager for Unilever and recently opened his own studio, Jan Hoekstra Design, in Rotterdam, which works on projects for such well-known Dutch brands as Droog Design, Henzo, HALO Nederland and Cor Unum. Hoekstra's work has won numerous awards and is in the permanent collections of the Stedelijk Museum Amsterdam, the Museum of Modern Art in New York, and the Chicago Museum of Architecture and Design. 96

Steven Holl was born in 1947 in Bremerton, Washington, USA. He graduated from the University of Washington and pursued architecture studies in Rome in 1970. In 1976 he joined the Architectural Association in London and established Steven Holl Architects in New York City. Considered one of America's most important architects, Steven Holl has realized cultural, civic, academic and residential projects both in the USA and internationally. The Kiasma Museum of Contemporary Art in Helsinki, Finland, (1998) is generally considered to be his masterpiece. He has been recognized with architecture's most prestigious awards and prizes. Steven Holl is a tenured professor at Columbia University's Graduate School of Architecture and Planning, has lectured and exhibited widely and has published numerous texts. 85

Richard Hutten was born in Zwollerkerspel, The Netherlands, in 1967. He graduated from the Academy of Industrial Design Eindhoven in 1991 and founded his own design studio, working on interior, furniture, product and exhibition design. Hutton is one of the most internationally successful Dutch designers, and a key exponent of Droog Design since 1993. Among his clients are Harvink, Moooi, Sawaya & Moroni, IDÉE Tokyo, S.M.A.K. Iceland, Pure-design Toronto, Centraal Museum Utrecht, Donna Karan New York, Maxfield Los Angeles, Karl Lagerfeld, KPN Telecom and HRH Queen Beatrix of the Netherlands. Additionally, Philippe Starck used some of his designs for the interiors of the Delano Hotel in Miami and the Mondrian Hotel in Los Angeles. Hutten's work has been widely published and exhibited worldwide. It is in the permanent collections of the Centraal Museum Utrecht, the Stedelijk Museum Amsterdam, the Vitra Design Museum in Weil am Rhein, Germany, and the San Francisco Museum of Modern Art, among others. 154

Giulio Iacchetti was born in Castelleone, Italy, in 1966 and has been an industrial designer since 1992. From 1998 to 2005 he collaborated with Matteo Ragni, with whom he founded the Aroundesign studio. In 2001 their project for a disposable fork, Moscardino, won the Compasso d'Oro ADI, and was later included in the permanent collection of the Museum of Modern Art, New York. Iacchetti lectures at various universities and design schools, both in Italy and abroad. Among his clients are Mandarina Duck, JVC, Guzzini, Sambonet, Coop Italia, Desalto, Bialetti, Breil, Pandora Design, Meritalia and Samsung. 95

IDEO, formed in 1991, is now the world's largest and most successful design and development firm, with offices in Palo Alto, San Francisco, Chicago, Boston, London, Munich and Shanghai. Its client list is an A–Z of leading international corporations, from Amtrak, BMW and Canon to Nike, Pepsi and Samsung. IDEO, which designs products, services, spaces, media and software-based interactions for organizations in the business, government, education and social sectors, has been independently ranked by global business leaders as one of the world's most innovative companies. 128

Inoda+Sveje was founded in 2000 in Copenhagen by Kyoko Inoda and Nils Sveje. Kyoko Inoda was born in Osaka, Japan, in 1971 and studied at the ISAD in Milan, Italy. Nils Sveje was born in Aarhus, Denmark, in 1969 and graduated from the Royal Academy of Fine Art in Copenhagen. The partnership has been based in Milan since 2003 and works for firms in Italy, Japan and Denmark, mostly on furniture, lighting and electrical appliances projects. 67

Massimo Iosa Ghini was born in Bologna, Italy, in 1959, studied architecture in Florence and graduated from the Polytechnic University of Milan. In 1985 he took part in the avant-garde movements of Italian design, creating illustrations, objects and interiors for the Bolidism group, which he founded, and also working with Ettore Sottsass of the Memphis group. During this period, he opened Studio Iosa Ghini in Bologna and began to work as an architect and designer. Iosa Ghini designs furniture, collections and objects, and handles art direction for leading Italian and international design firms. Iosa Ghini has held conferences and given lectures at various European universities, and has represented Italian design at a number of international symposia. 106, 168

James Irvine was born in London in 1958. He graduated with a degree in furniture design from the Royal College of Art, London, in 1984. Since then he has been based in Milan, Italy, where he runs his own design studio. He was a member of the Olivetti design studio until 1993 and a partner of Sottsass Associati until 1997. Today, Irvine's studio designs industrial products for companies such as Canon and Artemide, and among its clients are many furniture companies, including B&B Italia and Magis. 128, 186

Setsu and Shinobu Ito are both designers and architects. After Setsu graduated from Tsukuba University, and Shinobu from Tama Art University, both in Tokyo, Japan, they moved to Milan, Italy. They now work in Milan and Tokyo as consultants for import companies. Their work has been published and exhibited throughout Europe and Japan and has received several awards. Some pieces have been included in the permanent collections of museums of contemporary art in Munich, Germany, and Milan. Setsu is also a lecturer in Milan at the Domus Academy, the European Institute of Design and the Polytechnic University of Milan, and in Tokyo at the Tsukuba University and the Tama Art University. Shinobu also works as a graphic designer and is involved in marketing. 86

Toyo Ito is considered 'one of the world's most innovative and influential architects' (Designboom). He was born in Dailian, China, in 1941 and graduated from Tokyo University's Department of Architecture in 1965. He opened his own studio, Urban Robot, in 1971, changing the name in 1979 to Toyo Ito & Associates, Architects. He has won many awards, including the RIBA Royal Gold Medal in 2005, for his architectural projects, among the best known of which is the Sendai Mediatheque, Sendai, Japan. Ito holds a professorship at the Tokyo Women's University, is an honorary professor at the University of North London and has served as guest professor at Columbia University. 191

Marcia Iwatate was born in Georgetown, Washington DC, and began her career as an art director at Jun Co. for runway shows, advertisements and shop design. After working in New York from 1979 to 1980, she returned to Tokyo in 1983 and was creative director for Tokio Kumagai International (1988–90) and Intrigue, Mitsutomo (1989–91). In 1991, she began working in the area of restaurant concept development for various Tokyo restaurants. From 1992 to 2002, she was a creative director for Shunju Co. She is the author of five books, and her work has been widely published. 145

Emanuelle Jacques studied at ECAL, Lausanne, Switzerland, graduating in 2005. While still studying, she participated in several exhibitions, including the Milan Furniture Fair, and had the opportunity to design products for the Nanoo collection by Faser-Plast, BCV and Serralunga. She now works in Lausanne, where she has become a very promising talent on the Swiss design scene and has already received several awards. 43

Jacqueline Janssen studied at the Design Academy Eindhoven, The Netherlands, and worked as a textile designer for five years before joining Innofa, where she is now head designer. 199

Ulrika Jarl was born in Gothenburg, Sweden, in 1977 and left in 1996 after finishing her studies in media and advertising. After a few years of travelling and working, she continued her studies at the University of Brighton in England, where she studied three-dimensional crafts. She explores her love of natural forms and structures throughout her work as a designer–maker and finds inspiration in the patterns and structures of nature. Her products are made mainly from bone-china clay and slip, and she also uses various plastics to create similar translucent qualities. Ulrika Jarl Lighting and Homewares was launched in 2004. 144

Teemu Järvi was born in Espoo, Finland, in 1973 and graduated from the University of Art and Design, Helsinki, with an MA in furniture design. He has worked as a professional freelance furniture and interior designer since 1997 and, together with Heikki Ruoho, founded Design Studio Järvi & Ruoho, based in Helsinki, in 2003. He received grants from the Finnish Cultural Foundation in 2004 and 2005 and has won several awards. 74

Tahmineh Javanbakht, an artist and a painter originally from Iran, has produced many commissioned paintings. She graduated from the Art Center College of Design in Pasadena, California, in 1986, and later taught experimental painting there. Working in collaboration with Enrico Bressan, she formed Artecnica, and began to create a unique line of home accessories, from resin-made candleholders and frames to the popular jewel-toned bottle stoppers sold at upmarket retail stores worldwide. 129

Vicente Garcia Jiménez was born in Valencia, Spain, in 1978 and graduated in industrial design from the University of Experimental Sciences of Castellón de la Plana, in Spain. He then moved to Barcelona, where he worked with Santa & Cole developing lighting products and furniture, and soon produced his own prototypes. After moving to Milan, he met the architect Enrico Franzolini, and the two collaborated on a lighting collection for the company Karboxx. In 2005 Jiménez became the art Director of the lighting firm Fambuena and also set up his own studio in Udine, Italy, focussing on the design and industrial development of lighting nd furniture products, interiors and installations. His clients include Foscarini, Fambuena, Karboxx, Pallucco Italia and Tacchini. 146

Hella Jongerius was born in De Meern, The Netherlands, in 1963. She studied at the Design Academy Eindhoven from 1988 to 1993 and in 2000 she started her own company, JongeriusLab. Her work has been exhibited frequently and is also included in many museum collections. In 2003 she had a solo exhibition at London's Design Museum. 55, 93, 174, 188

Claudy Jongstra is a designer in felt who was born in 1963 in Roermond in The Netherlands. She has been commissioned by, or collaborated with, many furniture, interior and fashion designers, including Jasper Morrison, Ettore Sottsass, Steven Holl, Maarten Baas, John Galliano, Donna Karan and Christian Lacroix. Her work has won awards, been exhibited throughout Europe and in the USA and Mexico and is featured in the collections of several major museums and art institutions. 229

Patrick Jouin, born in 1967 in Mauves-sur-Loire, France, decided he wanted to become a designer when, at the age of seven, he spent the day at the Beaubourg Museum in Paris. He went on to graduate in industrial design from ENSCI/Les Ateliers in Paris in 1992. Jouin worked for Philippe Starck both at Tim Thom, Thomson Multimedia, then in Starck's own studio, at the same time producing his own lines for VIA, which gave him its Carte Blanche invitation (a grant to promote young designers) in 1998. In 1999 he opened his own studio, where he works as both a product designer and an interior architect, experimenting with materials and technology to push the boundaries of what is possible. 27, 60, 61, 150

Chris Kabel was born in 1975 and grew up in Bloemendaal in The Netherlands. After studying at Design Academy Eindhoven, he was an intern at CCD Architecturemoved in San Francisco. He now works from his own studio in Rotterdam. 66

Matthew Kavanagh was born in England in 1966 and studied product design at The South Bank University in London, graduating in 1988. He then worked as a product designer for Olivetti Computers in Italy, returning to the UK to set up his own lighting and product-design consultancy. In 1994 he began working for the first of several companies in the retail design sector, specializing in POP display, interiors, brand strategy and consumer psychology. He currently works with Astound, a group that focusses on design and strategy for retailers and brands. Along with Ashley Hall, he forms part of the Diplomat design partnership. 44, 94

Motomi Kawakami was born in Japan's Hyogo Prefecture in 1940 and studied at Tokyo National University of Fine Arts and Music, where he majored in industrial design. He received a master's degree from the same university in 1966, and in 1971 established the Kawakami Design Office. Kawakami specializes in product, interior, space and environmental design and has received numerous awards both in Japan and internationally. 195

Marcus Keichel *See* Julia Läufer

Lene Toni Kjeld lives and works in Kolding, Denmark. She received an MA in textile design from the Design School in Kolding in 2004. She designs wallpaper and carpets, and her work has been exhibited and published internationally. 207

Anthony Kleinepier was born and raised in Zoutelande, The Netherlands. After technical training, he studied at the Design Academy Eindhoven and at St. Joost Art Academy, Breda. He opened his own interior design business before graduating, and was also a member of CrashComfort, a collective concerned with design and entertainment. In 2002, he launched Soft Living Improvement Products under his Bone label. His work has been exhibited internationally. 24

Steven Koch graduated in product design from London's Central Saint Martins College of Art and Design. His company, Play Design, which is based in Cambridge, England, designs, manufactures and retails its own simple, functional and environmentally friendly products and also offers a design consultancy service to both business and public sectors. 186

Seppo Koho was born in 1976 and studied to be an interior designer at the Helsinki University of Art and Design, graduating in 1994. After working at several major architectural firms, Seppo started his own company in 1995 and has produced displays, interiors, furniture and lighting installations for a number of large Finnish companies. 164

Eero Koivisto was born in 1958 in Karlstad, Sweden, and is part of the Swedish design partnership Claesson Koivisto Rune (together with Mårten Claesson and Ola Rune), which was founded in 1995. As well as products, the trio design interiors and architecture. The partnership's furniture designs are produced by companies such as Asplund, Boffi, Cappellini, David Design, Dune, E&Y, Offecct and Swedese, among others. Eero Koivisto has held the position of artistic leader at the University College of Arts, Crafts and Design in Stockholm and has lectured at universities in Canada, Mexico and Norway as well as at other institutions in Sweden. 41

Harri Koskinen was born in Karstula, Finland, in 1970 and educated at the Lahti Design Institute, Finland (1989–93) and the University of Art and Design, Helsinki (1994–8). He has had solo exhibitions in Germany, Sweden, France, Italy, Japan, Singapore and the USA and has received numerous awards, including, in 2005, Design Plus, Frankfurt, and in 2004 the Compasso d'Oro, Milan, and Interior Innovation Award, Cologne. His work is in the collections of several prestigious museums, including the Museum of Modern Art, New York, the Museum of Art and Design, Helsinki, and the Chicago Athenaeum. 111

Kram/Weisshaar was founded by **Clemens Weisshaar** and **Reed Kram** in 2002 as a platform to integrate the multidisciplinary efforts of their offices. Based in Munich and Stockholm, the office engages in the design of spaces, products and media. Recent commissions include the concept, design, and execution of media installations and devices for Prada's new store in Beverly Hills, Los Angeles, an interactive installation for the Centre Georges Pompidou in Paris, data-visualization tools for the BMW group, master-plan and exhibition architecture for The Design Annual in Frankfurt, switch interfaces for Merten and furniture for Moroso and Classicon. The work of Kram/Weisshaar is part of the permanent collections of Die Neue Sammlung, Munich, the Vitra Design Museum in Weil am Rhein and the Centre Georges Pompidou in Paris. 65

Marc Krusin graduated from Leeds Metropolitan University, England, with a degree in furniture design. After placements with Fred Scott in London and George Sowden in Milan, he settled in Milan and worked in various studios, including Piero Lissoni's office, where he presently works as design manager and runs projects for clients such as Alessi, Kartell, Flos, Wella, Thonet, Fritz Hansen and Knoll. In 1998, Krusin co-founded the Milan-based group Codice 31 with five other designers from varying ethnic backgrounds. The group made its debut at the 1998 Milan Furniture Fair's 'Salone Satellite' exhibition for young designers, and in successive years expanded, and collaborated with a number of prestigious companies. In 2004, Krusin opened his own company, Klay, specializing in natural objects from world cultures. 33, 205

Mikko Laakkonen was born in Espoo, Finland, in 1974 and trained as a musical-instrument maker before studying design at the University of Arts and Design, Helsinki. After several years of working for design companies in Finland, he founded his own studio in 2004 in Helsinki. Laakkonen works with various international clients, mainly in the fields of furniture design, product design and interior architecture. He has participated in several international and local exhibitions, including the Saunabus exhibition in 2002, the Rehti exhibitions at the Milan Furniture Fair in 2005 and 2006, and Helsinki Design Week in 2005. 95

Gregory Lacoua was born in Tours, France, in 1979 and is a student of design at ENSCI/Les Ateliers in Paris. His mat/stool A Needle and Thread was developed as a student project in collaboration with VIA. 88

Marta Laudani and Marco Romanelli, Italian architects, have been collaborating since 1986 in Rome and Milan. They work in the fields of design (for Driade, Montina, Laboratorio Pesaro, O Luce, Salviati among others), interiors and exhibition design. Recently, they designed the restoration of the Museum of Roman Civilization for Fiat Engineering in Rome. Laudani teaches history of design on the masters programme for curators of museums of contemporary art and architecture at La Sapienza University in Rome; Marco Romanelli was director of the journal *Domus* from 1986 to 1994, and since 1995 he has held this position for *Abitare*. 78, 117, 176, 181, 215

Julia Läufer was born in 1968 in Freiburg, Germany, and studied fashion and textile design in Trier, London and Berlin, graduating from Humboldt Universität in 1994. **Marcus Keichel**, born in 1967 in Frankfurt, Germany, studied mechanical engineering in Cologne, and then industrial design and architecture at the Universität der Künst Berlin, and graduated in 1995. Läufer and Keichel formed a partnership and opened their own studio in 2000, specializing in product design and corporate design. Their work has received two Red Dot awards for product design. 60

Giovanni Levanti was born in Palermo, Italy, in 1956. After receiving his degree in architecture in 1983, he moved to Milan to attend the Domus Academy, and from 1985 to 1990 collaborated with Andrea Branzi, while also (from 1986) working as an independent industrial designer. Levanti's clients include Cassina, Domodinamica, Edra, Foscarini, Marutomi, Memphis, Pallucco, Salviati, Twergi-Alessi and Serafino Zani, and his work has won or been shortlisted for several international awards. He is also a visiting professor at the Domus Academy. 75

Arik Levy was born in Tel-Aviv and received a degree in industrial design from Art Center Europe in Vevey, Switzerland, in 1991. He is the creative director of and a partner in Ldesign, Paris. His work has been shown in many exhibitions in museums, alternative spaces, galleries and fairs. He has received many awards, and his work is in the permanent collections of various museums. His clients include Vitra, Visplay, Ligne Roset, Desalto, Baleri Italia, Gaia & Gino, Cinna, Dietiker, Magis, Serralunga, Ansorg, Belux, Lafayette, L'Oréal and Lampert. 35, 85, 171, 184, 189

Lievore, Altherr, Molina was founded in 1991 by Alberto Lievore, Jeanette Altherr and Manel Molina. The partnership is currently working for firms in Spain, Germany and Italy, most notably on furniture projects but also developing interior, packaging and product design and on consultancy and art direction projects. The partnership has received numerous awards both in Spain and abroad and its work has been exhibited in Europe, the USA and Japan. Barcelona. Its designs are featured regularly in Spanish and international trade magazines. 34, 35, 44, 57, 69

Piero Lissoni was born in 1956 and studied architecture at the Polytechnic University of Milan, Italy, then worked for G14 studio, Molteni and Lema. In 1984 he co-founded a company designing product, graphic and architectural projects. Since 1986 he has worked with Boffi Cucine, Porro, Living Design and Matteograssi. He became art director for Lema, then for Cappellini and in 1998 began working with Benetton. 56, 69, 76, 77, 132

Aron Losonczi was born in Szolnok, Hungary, in 1977 and studied architecture and engineering at the Technical University in Budapest, graduating in 2001, and at the Royal Institute of Technology and the Royal Institute of Art, both in Stockholm, Sweden. After working for a number of architectural firms, he opened his own practice in 2002. 212

Ross Lovegrove was born in 1958 in Cardiff, Wales. He studied industrial design at Manchester Polytechnic in England, then gained an MA in design from the Royal College of Art in London. In the early 1980s Lovegrove worked in Germany as a designer for Frog Design, on projects such as Sony Walkman and Apple Computers, before moving to Paris as a consultant to Knoll International. He was invited to join the Atelier de Nîmes – along with Jean Nouvel and Phillipe Starck – providing design consultancy for companies such as Cacharel, Louis Vuitton, Hermès and DuPont. Since he returned to London in 1988, his clients have included British Airways, Kartell, Cappellini, Idee, Moroso, Loom, Driade, Peugeot, Apple Computers, Olympus Cameras, Luceplan, Tag Heuer, Hackman, Japan Airlines and Toyo Ito Architects, Japan. His work has won numerous international awards, been extensively published and exhibited across the world and is held in the permanent collections of the Museum of Modern Art, New York, and London's Design Museum. 29, 134, 143

Margit Lukács *See* Persijn Broersen

Ane Lykke graduated from the Danish School of Design, Copenhagen, in 1996, and also attended the Kawashima Textile School in Japan in 1998. She has worked in studios in Rome and Tokyo and has undertaken commissions for clients such as Bang & Olufsen (1996–8), Trapholt Art Museum (2006–7) and Stilleben, Copenhagen (2002–3, 2006). Her work has appeared in nine collective design exhibitions, and in 2006 there was a solo exhibition of her work, entitled 'Mind the Gap' at The Danish Design Centre. 202–3, 222

Vico Magistretti was born in 1920 in Milan and obtained his degree in architecture in 1945. In 1956 he was one of the founders of ADI (Associazione per il Disegno Industriale, the Industrial Design Association) and in 1960 he began to work at mass furniture design, creating the first plastic chair, which was first manufactured in 1967. Eighty per cent of all Magistretti's designs are still in production. His works have been displayed in the most relevant design exhibitions in Italy, Europe, USA and Japan, and are part of the permanent collections of the most prestigious museums worldwide, including the Museum of Modern Art in New York, whose collection includes twelve of his design pieces. Magistretti has been awarded with many important prizes and titles and has lectured in design worldwide. 57

Cecilie Manz was born in Denmark in 1972 and studied furniture and product design at the Danish Design School, Copenhagen, and at the University of Art and Design, Helsinki, Finland. She opened her own studio in Copenhagen in 1998. Her work has been shown at the Museum of Modern Art, New York, the Vitra Design Museum, Basel, Switzerland, the Danish Museum of Art and Design, Copenhagen, and the International Museum of Applied Art, Turin, Italy, as well as in exhibitions in Japan and throughout Europe. 151

Javier Mariscal opened the Mariscal Studio in 1990. He recently designed the Hotel Domine opposite the Guggenheim in Bilbao, Spain, and is also working on a range of graphic, audio-visual and editorial projects there. 37, 52

Nani Marquina was born in Barcelona, Spain, in 1952 and studied industrial design at the Massana School there. In 1987 she created her own brand, specializing in the design of textiles for the home, with a particular focus on rugs. Marquina's designs have been featured in a number of group exhibitions worldwide and have received several awards. Her Topissimo collection won both the Red Dot award and the Chicago Atheneaum's Good Design Award in 2003. 223, 230

Chris Martin, born in 1970, graduated from the Royal College of Art in London in 1995 and now designs furniture, lighting and other products. He has worked as an assistant to Jasper Morrison and as a designer at the Sandellsandberg design partnership in Stockholm, Sweden. He has also lectured at Beckmans School of Design in Stockholm and was a finalist in a snow-sculpture competition held in the north of Sweden. His work has been recognized with the Excellent Swedish Design award. 87

Simona Marzoli *See* Fabrizio Bertero

Jean-Marie Massaud was born in 1966 in Toulouse, France, and graduated from ENSCI/Les Ateliers, Paris, in 1990. His collaboration with Marc Berthier and his work on town-planning projects inspires him to fuse design and architecture, and he is involved in many different aspects of design, from furniture through to industrial products. Clients have included Yamaha Offshore and Renault, Italian furniture companies Cappellini and Cassina, and luxury brands Armani and Baccarat. 51, 52, 72

Shoichiro Matsuoka was born in Japan and worked for International Industrial Design Inc. before joining Sony in 1982. He has designed a wide range of products, including a TV and a car audio system. 113

Franz Maurer, born in 1963, lives and works in Vienna and in Waldviertel, Austria. He founded his own design studio in 1989 and specializes in product design, presentations and the design of light and living objects. Maurer works mostly with industrial materials, and his clients include Artek, Finland, and Artificial, Ingo Maurer and Raumgestalt (all in Germany). He has won several prizes, including the Design Plus Award and the Red Dot award, both in 2005. 172

Ingo Maurer was born in 1932 and began to design exceptional lighting and lighting systems in 1966. His company has produced and distributed innovative lighting products worldwide for the past forty years. He and his team develop concepts and spectacular one-offs for private and public buildings, such as Bulb (1966), the low-voltage halogen system YaYaHo (1984) and Lucellino (1992), a winged light bulb, which are among his best-known designs. A number of his works have been included in the permanent design collection of the Museum of Modern Art in New York. 144, 167

Mario Mazzer was born in 1955 and graduated in architecture from the Polytechnic of Milan, Italy, in 1978 and gained a diploma in industrial design from the Polytechnic School of Design in Milan the following year. During the 1970s he gained experience working with masters of Italian design, and in 1980 opened his own studio in Conegliano, Veneto. Mazzer is involved in projects from architecture to industrial and furniture design and has won a number of prizes. He contributes to the cultural debate on design in conventions and conferences and by showing his work in events to promote products made in Italy. A monograph was published about him in 2001 by Biblioteca dell'Immagine. 47

Alberto Meda was born in Lenno Tremezzina (Como), Italy, in 1945 and now works and lives in Milan. He received a master's degree in mechanical engineering from the Polytechnic of Milan in 1969 and from 1973 worked as technical manager of Kartell, responsible for product development. Since 1979 he has worked as a freelance industrial designer. Clients include Alias, Alessi, Arabia-Finland, Ansaldo Sistemi Industriali, Cinelli, Colombodesign, Italtel Telematica, JcDecaux, Mandarina Duck, Ideal Standard, Luceplan, Kartell, Omron Japan, Philips, Vitra and Olivetti. Meda has lectured on industrial technology at the Domus Academy, the Polytechnic of Milan and at IUAV of Venice, as well as at conferences in Europe, Japan and the USA, and is a member of the board of Designlabor Bremerhaven. His work has received several awards, including the Compasso d'Oro (1989 and 1994) and the Bundespreis

Produktdesign (2000). Several of his products are in the collection of the Museum of Modern Art in New York. 48, 128

Yael Mer was born in Israel in 1976. She received a BA in design from the Academy of Art and Design, Jerusalem, in 2002, and is currently completing an MA in design from the Royal College of Art, London. In 2003–4 she worked on display systems and digital printing for Campus, and was artist in residence at the Japan Mino paper-art village project. She has been awarded prizes for her food packaging and other designs, and her work has been exhibited internationally since 2002. 98

Lucy Merchant was born in 1974 and graduated with a degree in fine art from the Kingston School of Art, London, in 1997. In 2004 she established the Lucy Merchant Design Studio, where, focussing on the relationships between art, architecture and design, she has been developing projects and design collections for leading manufacturers and clients. Merchant's designs have been published and exhibited internationally. 50, 106

Luca Milano studied architecture at Turin Polytechnic and design at the Domus Academy in Milan, Italy. After graduation, he began work as a furniture designer and manufacturer and founded his own company, La Maison. He has designed and furnished the interiors of hundreds of houses and shops in Piemonte, Italy. Since 2000 he has collaborated with Inci Mutlu as Mutlu+Milano Design Studio. 137

Jeff Miller has his own design studio in New York, and acts as a consultant for product and furniture manufacturers. He was previously vice-president of design at ECCO Design, New York, where for thirteen years he was involved in the development of consumer electronics and other products, medical and sports equipment, and residential and office furniture. He has received numerous design awards, and his work has been published and exhibited internationally. Miller holds several patents and has been invited to lecture on his design accomplishments at prominent design schools and institutions worldwide. 51, 77

Rosita Missoni has spent nearly fifty years at the helm of Missoni with her husband, Ottavio Missoni, gradually creating a distinctive international style. In 1997 she stepped aside so that her three children could take over her role in the company, and then devoted herself to developing collections for the Missoni Home range. The collection has already won many important awards. 202, 216

Vika Mitrichenka was born in Minsk, Belarus, in 1972. He studied in Amsterdam, The Netherlands, at the Gerrit Rietveld Academy (2000–4) and the Sandberg Instituut (2004–5) and is currently continuing his studies at the Rijksakademie van Beeldende Kunsten. His work has been exhibited in solo and group shows at various museums and galleries in The Netherlands, Belgium and Germany, is in collections of the Stedelijk Museum and the Nederlandsche Bank, Amsterdam, and has been featured in several magazines, including *Frame*. 190

Modoloco Design Milano was founded in 2001 with a goal of creating an open space for experiments and research in which culture and different experiences could be realized in the form of architecture, design, and communicative graphics. Modoloco collaborates with important Italian and international companies, including Firme di Vetro, Muranodue, Talleruno, Celda Iluminacion, International View, Yamakawa and Fondermetal. 151

Ito Morabito was born in Marseilles, France, in 1977. He left design school after only one year and worked for an architect and then for the shoe designer Roger Vivier. In 2002 Ora-ïto, Morabito's brand name, received an award for an aluminium Heineken bottle. He has also designed the OGO oxygen-enriched water bottle and adidas and Joop! perfume bottles, and has devised advertising campaigns for Alain Mikli, Levi's and OGO. Other clients include Cappellini, Magis and Artemide. 161

Lauren Ruth Moriarty graduated from the Loughborough University School of Art and Design with a degree in multimedia textiles, and received her MA in industrial design from Central Saint Martins College of Art and Design, in London. She is a three-dimensional textile designer who focusses on experimentation with materials such as rubber and plastics, not usually incorporated in

textiles, and working with unusual industrial processes. She designs lighting, cushions, interior cubes and home accessories. 220

Ulf Moritz graduated as a textile designer from Krefelds Textilingenieurschule in Germany in 1960. He has been creating fabrics in his own design studio in Amsterdam for Sahco Hesslein since 1970. In 1986 he introduced his own fabric collection, Ulf Moritz by Sahco Hesslein, to the market. He is currently art director for Danskina and in 2006 will launch his carpet collection, Ulf Moritz by Vorwerk. His textile and interior designs have been exhibited internationally and he was recently given the honorary title Royal Designer for Industry by the Royal Society of Arts in London. 224

Jasper Morrison was born in London in 1959. He studied at Kingston Polytechnic and undertook post-graduate work at the Royal College of Art and at the Hochschule der Künste, Berlin. In 1986 he set up his Office for Design in London, since when he has worked for Alessi, Alias, Cappellini, Flos, Magis, SCP, Rosenthal and Vitra. In 1995 his office was awarded the contract to design the new Hannover tram for Expo 2000. Recent projects have included furniture for the Tate Modern in London. 22, 38, 66, 67, 96

Mosley meets Wilcox was formed in east London by Steve Mosley and Dominic Wilcox after both graduated from the Royal College of Art in 2002. The partnership soon became well known for products such as the War Bowl, Dip Lamp and Honesty Stamp distributed by Thorsten Van Elten. More recently, after a chance meeting at a photo shoot in east London, they began a collaboration with iconic music photographer Mick Rock, forming the project Mosley meets Wilcox meets Rock. The collection was launched at the Hanbury Gallery during London Design Week in 2004, and later a limited-edition collection featuring images of David Bowie was sold exclusively through Paul Smith stores worldwide. 182

Inci Mutlu graduated in industrial design from Middle East Technical University, Ankara, Turkey, in 1994, and went on to receive an MA in design. Between 1996 and 2000 she worked as a freelance designer in Istanbul, later moving to Italy to work for Isao Hosoe Design Studio before collaborating with Luca Milano as Mutlu+Milano Design Studio. Her work has received several awards, including first prize in the Viridian Design Competition (USA) in 2001 and, more recently, the Best Designer and Best Bathroom Design awards of Elle Décor (Istanbul) in 2005. Her clients include Vitra, Nurus, Koziol, Guzzini, S.Pellegrino, Droog Design, Tribu, Jongform, Tuna-Girsberger, Winston and Restonic. 137

Khashayar Naimanan was born in London in 1976 and is of Persian–Japanese origin. He studied at Central Saint Martins in London (1996–9) and then at the Royal College of Art, during which time he won the Blueprint Competition organized by the Vitra Design Museum (2001) and began working with Porzellan Manufaktur Nymphenburg. This collaboration continues, and he is now one of four designers working on their contemporary collection of tableware. Khashayar lives and works in London and Munich. 187

Oki Sato Nendo, who was born in Canada in 1977, won numerous awards as a student and received his MA in architecture from Waseda University in Tokyo in 2002. Upon graduating, he founded his own company, which undertakes projects in architecture and interior design, and also furniture and graphic design, for Japanese and European clients. It has received numerous international awards, such as the Tokyo Designers Block 2002 and 2003, the Good Design Award 2004 and the JCD Design Award 2005. 26, 27, 83, 154

Marc Newson's creations range from household objects, furniture, restaurants and watches to aircraft interiors. Australian-born Newson has won acclaim internationally, as well as a clientele that includes Flos, Cappellini, Magis, Nike, Alessi, Samsonite, Ideal Standard, Qantas and Ford. His works are in the collections of the Museum of Modern Art in New York, the Design Museum in London and the Musée National d'Art Moderne and the Centre Georges Pompidou in Paris. 109

Masami Nitta studied graphic design at the Camberwell College of Art in London and joined Sony in 2001. She has designed a range of graphics and packaging. 113

Nya Nordiska is a renowned textile company in Germany, founded by Heinz Röntgen in 1964. Nya Nordiska has won a large number of international design awards. 210–11

Patrick Norguet, who lives and works in Paris, has a vocational qualification in industrial draughtsmanship, lathing and milling, and is a production-technology engineer. He has been commissioned by several furniture manufacturers, such as Cappellini, Livit and Moroso. His earlier works include the Boson, Little Apollo and Apollo chairs, and What's Up and Lex sofas for Artifort. He has received a number of awards and distinctions. 53

Norway Says, a design studio based in Oslo was founded in 2002 by Espen Voll, Andreas Engesvik and Torbjørn Anderssen after three years of exhibiting at fairs such as the Milan Furniture Fair, Designers Block in London and the Stockholm Furniture Fair. Today the company works with various international clients in the areas of product, furniture, installation and interior design. 49, 127

Marie O'Connor is a designer, illustrator and fashion consultant who works on collaborative and independent projects and has built a diverse portfolio of work across a wide range of disciplines. She receives commissions to produce illustrations for many magazines and advertising campaigns and is a member of Peepshow, a London collective. Her work has been exhibited in London, Hamburg and New York, and in 2005 she had a solo show at The Lighthouse, Centre for Design & Architecture, in Glasgow, Scotland, featuring textile work and short films, which were subsequently shown at the Institute for Contemporary Arts and the Victoria & Albert Museum, both in London. 108

Frank Oehring is a sculptor who has lived and worked in Berlin since 1967. His work crosses the boundary between free and applied art and design. He has designed stage rooms in cooperation with composers, choreographers and actors, and has received numerous commissions for art in public spaces, wall reliefs and outdoor, kinetic and lighting sculptures. He produced the design for the complete information and guide-system inside the International Congress Centre in Berlin. 158

Reiko Okamoto was born in Tokyo and graduated from Musashino Art University there. She designs tableware and lighting fixtures and has been employed by the Noritake Company (1986–9) and Sazaby. (1997–8), among others. In 1994, she began her collaboration with Marcia Iwatate. From 1998 to 2002 she was a creative director for Shunju Co. 145

Omlet was created by James Tuthill, Johannes Paul, Simon Nicholls and William Windham, who met while studying at the Royal College of Art, London. They launched the Eglu chicken house soon after as a way for people to produce and collect eggs at home. 129

Ora-ïto *See* Ito Morabito

Dr Margaret Orth received her PhD in media arts and sciences from the Massachusetts Institute of Technology Media Lab in 2001. Her work includes patents, research, publications and design in new physical interfaces, wearable computing, electronic textiles and interactive musical instruments. Orth is a pioneer in the developing field of electronic textiles, interactive fashions, wearable computing, and interface design. Her groundbreaking work in electronic textiles has appeared in numerous publications and has been widely exhibited in international venues, including the National Textile Museum in Washington DC and the Stedelijk Museum Amsterdam. 116

Jay Osgerby *See* Barber Osgerby

Kensaku Oshiro was born in Okinawa, Japan, in 1977 and studied in Italy at the Polytechnic of Milan, graduating in 1999. He is currently working at Lissoni Associates, as well as collaborating with other companies as an independent designer. 39, 111

Clare Page *See* Committee

Satyendra Pakhalé was born in India in 1967. After completing his MA in design in India and a degree in advanced product design at the Art Center College of Design in Vevey, Switzerland, he worked on major assignments such as the interior for the Pangéa Concept-car, developed in collaboration with Renault and Phillips. In 1998 he set up his own design practice in Amsterdam, and since then has been working on his own projects and on designs for a diverse group of clients, including Alessi, Cappellini, Colombo Design, De Vecchi, Erreti, Magis, Moroso and RSVP (Italy), Cor Unum (NL), C-Sam (USA/UK), and Material ConneXion (USA). Pakhalé's work has recently been the subject of a book and has been shown in two solo exhibitions, one in the Stedelijk Museum Amsterdam and another at the Otto Gallery, Bologna. 107

Paolo Pallucco and Mireille Rivier are a husband and wife design team based in Rome, Italy. 50

Ludovica and Roberto Palomba have been working together since 1994. After graduating from the University of Rome, Italy, they founded architecture and design practice Palomba Serafini Associati, based in Verona. Their work has been exhibited internationally and their product designs have received numerous awards. Among their clients are Bosa, Crassevig, Flaminia, Iris, Kos, Schiffini, Zucchetti and Tubes. Roberto has been a professor in the Industrial Design Department of the Polytechnic University of Milan since 2003. 25, 81, 106, 131, 134, 135

Andrea Panto *See* Fabrizio Bertero

Donata Parruccini was born in Varedo, Italy, in 1966 and studied industrial design at ISIA in Florence with Jonathan De Pas. From 1994 to 1997 she worked with Andrea Branzi and then as a freelance designer. Her designs are manufactured by Alessi, Morellato Pandora Design and RSVP. In 2004 she taught at the European Institute of Design in Milan. Her work has appeared in a number of group exhibitions, including '1950–2000: Theater of Italian Creativity' at the Dia Center in New York (2003) and the 'Caiazza Memorial Challenge' (2005). 178

Tim Parsons was brought up in Wiltshire, England, and studied industrial design in Bournemouth and Teesside before attending the Royal College of Art in London from 1998 to 2000. After graduating with a degree in product design, he began selling his own designs to London retailers and was commissioned to undertake a research project for the Helen Hamlyn Research Centre, dedicated to the promotion of socially inclusive design, based at the Royal College of Art. In 2002 he moved to Manchester, where he now combines teaching at Manchester Metropolitan University with running a small studio. He has collaborated on design projects with internationally renowned firms such as Matali Crasset and Vogt & Weizenegger and has led teaching workshops in Lund, Sweden and Reims, France. His products have been manufactured by Innermost, Thorsten Van Elten and The Berlin Institute for the Blind, and his work has been exhibited widely, including at the Design Museum, London. Parson's recent work has focussed on experiments with pewter for the Sheffield-based manufacturer A. R.Wentworth. 173

Vittorio Passaro was born in Montella, Italy, and studied cabinetmaking at Meda, near Milan, and sculpture at Fine Arts Academy Brera, Milan. His first exhibition was in 1992 at the Gallery Via Farini in Milan. In 1994 he went to Los Angeles for six months to work on a video-art project, and then continued his artistic research in Paris where he stayed for three years. In 2000 he began working on design projects with his brother Daniele: together they developed an altar in the Church of San Francesco in Folloni, Montella, Italy. Since 2002 he has been collaborating with Patricia Urquiola in Milan. 185

Eva Paster was born in Munich, Germany, in 1971 and studied industrial design at the University of Applied Sciences, Munich. In 1997, while still studying, she co-founded the company Neuland Industriedesign. In collaboration with **Michael Geldmacher**, she designs furniture for MDF Italia, Interlübke, Cor and Nils Holger Moormann. She has been a lecturer at the University of Applied Sciences since 2002. 33

Pearson Lloyd is a multidisciplinary design studio based in Central London. The studio's work focusses on design for manufacture, strategic brand development and research in the fields of furniture, transport design and the public realm. Pearson Lloyd has won numerous awards throughout Europe and America for its work, including eleven internationally acclaimed awards for Virgin Atlantic Airway's new Upper Class Suite. 19

Terri Pecora was born in California in 1958 and studied fashion illustration at Art Center College of Design in Pasadena. In 1988 she transferred to the product design masters programme at Domus Academy in Milan, Italy, and in 1991 established her own studio. Her work includes interior design, furniture design, exhibition installations, art direction and graphics, fashion accessories, products for the home, wall coverings and bathroom fittings. Among her clients are adidas Eyewear, Art & Cuoio, Bisazza, BRF, Marco Bicego, Dom Ceramica, Edra, Esprit Eyewear, Interflex, Flou, Montblanc, Persol Eyewear, Plumcake Kids, Prénatal, Simas, Swatch, and Zanotta. Pecora has taught at Domus Academy, the European Institute of Design, at the University La Sapienza of Rome and, since 1998, at the Polytechnic University of Milan. 139

Gaetano Pesce is an architect–designer, based in New York, who has undertaken diverse commissions in architecture, urban planning, interior and exhibition design, industrial design and publishing. In more than forty years of practice, Pesce has worked on public and private projects in the USA, Europe, Latin America and Asia, from residences to gardens and corporate offices. In 1996, he was honoured with both a comprehensive career retrospective at the Centre Georges Pompidou in Paris and the publication of an exhibition catalogue of his work. In 1993 he was the recipient of the prestigious Chrysler Award for Innovation and Design. 149

Gabriele Pezzini was born 1963 in Charleroi, Belgium, and studied at the Institute of Industrial Design in Florence, Italy. He joined the international design team of the French company Allibert in 1991, and later became design manager, working with the company until 1997. In 1999 he opened his own studio in Milan, where he has continued his research work, as well as freelancing for new companies in different sectors. He has also taught as a visiting lecturer at the Ecole des Beaux-Arts de Saint-Etienne in France, at the Rhode Island School of Design, in Providence, and at the Polytechnic University of Milan. 119

Christophe Pillet is a French designer who studied at the Domus Academy in Milan, Italy, and after graduating with a master's degree in design, worked with Philippe Starck in Paris from 1988 to 1993. Since setting up his own company, he has worked on projects ranging from architecture and interior design to industrial design, furniture and fashion. He won the French Designer of the Year award in 1994. 39, 79, 89, 142

Russell Pinch was born in 1973 and studied at Ravensbourne College of Design, London. After graduating, he worked as Sir Terence Conran's design assistant and in 1995 he became a senior product designer for the Conran Group. Here he was responsible for developing a diverse range of products for the Conran shops and restaurants and many of the designs for the Conran Collection, Conran's benchmark homeware collection. After five years with Conran, Russell co-founded The Nest, a multidisciplinary brand design agency with clients including British Airways, MFI, WH Smith, Rip Curl and Selfridges. In 2004 a return to furniture design beckoned and Pinch, a furniture, product and interior design company, was born. The first collection of furniture was launched at 100% Design, London, to great acclaim. Pinch was awarded the Blueprint/100% Design Best Newcomer award and the Design & Decoration Furniture Award 2005. 82

Franco Poli was born in Padua, Italy, in 1950. On completing his studies in Venice (IUAV and International University of Art) in 1974, he began working as a designer in Florence, Venice and Verona. He is a firm defender of the originality and independence of Italian design, and in the 1990s founded the AD Veneto associations and the Italian Chamber of Design. He was a professor of design at the historic Academy of Fine Arts in Venice until 1996, and is currently visiting professor at the Polytechnic of Milan and at the UIA in Florence. He has been responsible for many publications and has taken part in numerous exhibitions, for which he has received awards, both in Italy and abroad. Poli currently collaborates with a number of companies, most notably Matteograssi. 29.

Bertjan Pot was born in Nieuwleusen, The Netherlands, in 1975 and graduated from the Design Academy Eindhoven in 1998. From 1999 to 2003 he collaborated with Daniel White under the name Monkey Boys. In addition to their own collection of furniture and lighting, the two designers also created products for Moooi, Goods and Lietmotiv. Pot's work has been exhibited in numerous venues in The Netherlands and also in other parts of Europe and in Japan. His work is also in the collections of the Museum Booijmans Van Beuningen, Rotterdam, and the Victoria & Albert Museum in London. 30–1, 74, 165

Itay Potash was born in Israel in 1978 and graduated in industrial design from Bezalel Academy of Art and Design, Jerusalem, in 2004. During 2004–5 he worked as a user-experience designer at SAP Labs in Israel and in 2005 established his own studio in Gezer, between Tel-Aviv and Jerusalem. His work has appeared in numerous design publications and in several exhibitions. In 2004 his Flat Mode sewing machine, which utilizes Bluetooth technology, won first prize in the Manufacturers Association of Israel competition for excellence in industrial design. 108

rAndom International was founded in 2002 by Flo Ortkrass, Stuart Wood and Hannes Koch, all graduates of the Royal College of Art, London. In addition to developing a number of self-initiated technologies, products and installations, part of rAndom has grown into Operation:Schoener, a design and creative technology consultancy that has a variety of clients, including London-based art curators Artwise, Nokia and TribeArt Projects. Their work has received various awards, including the iF (International Forum Design) concept award and two others for their PixelRoller projects in 2006. 212

Ransmeier and Floyd design studio was founded by Gwendolyn Floyd and Leon Ransmeier in The Netherlands in 2003. The studio designs products and furniture and handles interior design projects. Floyd and Ransmeier met at the Rhode Island School of Design, Providence, from which Ransmeier graduated in 2001, and subsequently moved to The Netherlands, where Floyd graduated from the Design Academy Eindhoven after working as an intern under Sam Hecht at Industrial Facility. The work of Ransmeier and Floyd has received numerous international awards and they have an international client base. They have been selected for the 2006 Design Triennial at the Cooper-Hewitt National Design Museum in New York. 94, 128

Karim Rashid, who is half English, half Egyptian, was born in Cairo, Egypt, in 1960 and raised mostly in Canada. He studied industrial design at Carleton University in Ottawa and pursued graduate design studies in Naples, Italy, with Ettore Sottsass and others before moving to Milan to spend a year at the Rodolfo Bonetto Studio. He then worked for seven years with KAN Industrial Designers in Canada before opening his own practice in New York City in 1993. Rashid has worked for numerous clients globally, won many awards for products, packaging and restaurant interiors, and has more than seventy objects in permanent collections. His design work has been exhibited at major museums and galleries throughout North America and in London, Milan, Hamburg, Seoul, Tokyo, Austria and The Netherlands. Rashid has been a juror for several international competitions, a contributing writer to design periodicals, lectures internationally and has held posts in industrial design at prominent design institutions in the USA and Canada. 174

Gerhard Reichert was born in Ravensburg, Germany, in 1962 and attained a degree in mechanical engineering, followed by another in integral design at the SADK, Stuttgart, Germany. He has worked for the Siemens Design Centre and, together with the Fraunhofer Institute, has developed new modular car systems for Skoda. From 1993 to 1997 he was project manager at MH Design Engineering in Switzerland and Germany, and he became project manager at Studio De Lucchi in Milan, Italy, in 1997. Reichert later founded his own studio, and since 2004 has been professor of design at the Aachen University of Applied Sciences in Germany. He designs interiors, office furniture, lamps and industrial products and collaborates with numerous companies, including Artemide, Interstuhl, Nava and Sitag. His work has received many awards. 92

Harry Richardson *See* Committee

Garth Roberts, who trained as an industrial designer, now designs furniture, products, housewares and interiors and has collaborated with Saint-Gobain Glass, Fasem, Zanotta and Donna Karan New York. He has won awards for his product and furniture design and his work has been shown in the USA, Japan and in France. Currently Garth's focus is on developing his studio with contact points in New York, Milan and Berlin. 82

Vibeke Rohland, textile artist, is based in Copenhagen, Denmark. She graduated from the Danish Design School in Copenhagen, where she studied textiles, in 1986, and from 1987 to 1989 was an assistant for M. Eliakim in Paris, and hand-painted textiles for Parisian *haute couture* and interior *couture*. She is a designer for Georg Jensen Damask, and has created the Crossings textile collection for the bedroom, bathroom and kitchen, which will be launched in autumn 2007. Among her clients are IKEA, Bodum, Kvist Furniture, Esprit New York and Agnes B. She has received numerous grants and her work has been exhibited internationally. 209

Frederik Roijé was born in 1978 in Goor, The Netherlands. He graduated from the Design Academy Eindhoven in 2001 and opened his own design studio in Amsterdam in 2002, specializing in industrial product design. His work includes cocktail glasses designed for Bacardi-Martini Nederland (2002), Spineless Lamps for Cor Unum (2002) and ceramic products for Droog Design (2004). In 2005 he won the Nederlandse Design Prijzen. 160

Marco Romanelli *See* Marta Laudani

Phillip Rose graduated from Northumbria University, Newcastle, England, in 1987. He began work as a product designer at Sony in 1990, and has become the senior manager of Product Design and Human Interface Design. He has produced a diverse portfolio of designs, including personal and professional audio products and home entertainment equipment. 113

Galya Rosenfeld studied at Bezalel Academy of Art and Design in Jerusalem, and founded her own design studio after graduating in 2001. She works at the intersection of fashion, craft, design and art. Her work is part of the permanent collections of the Costume Institute of the Metropolitan Museum of Art in New York, and she has been commissioned to produce a piece for the Museum of Modern Art in New York. She teaches, lectures and exhibits her work internationally. 231

Diego Rossi and Raffaele Tedesco are Italian architects and designers, both born in 1976 and both graduates in industrial design from the Polytechnic of Milan's Department of Architecture. The two have done a great deal of research on the use of renewable and alternative energy sources for lighting in association with 3M and German company Bomin Solar Research. Since 2001 Rossi and Tedesco have collaborated with Alberto Meda and Paolo Rizzatto on designs for Luceplan which have garnered many awards, most notably the Design Plus. 156

Adrien Rovero, born in 1981, lives and works in Lausanne, Switzerland. He received a BA in product design from the Ecole Cantonale d'Art de Lausanne in 2004, and won a number of awards while working on his degree. Rovero has his own studio, Inout. 200

Tomek Rygalik grew up in Poland and studied architecture in Lódz and industrial design at the Pratt Institute in New York. He then worked with several design consultancies in New York before returning to study in London on the Royal College of Arts Design Products postgraduate programme. Since graduating in 2005 he has run his own design practice and also works part time as a research associate at the RCA. Tomek has been awarded with numerous prizes and awards, including First Prize Award in the 2006 International Bombay Sapphire Martini Glass Design Competition, BSI Environmental Design Award 2005, and Rosenthal Design Award 2004. Two of his furniture pieces were part of the British Council's Talent/Talento selection in 2005. Over the last few years his work has been exhibited in London, Milan, Munich, New York, Tokyo, Poznan and Valencia. 68

Toshihiko Sakai was born in Kochi Prefecture, Japan, in 1964 and studied design at Tokyo Zokei University (1989–91). He founded his own studio, Sakai Design Associates, in 1992, working in the fields of product, interior and furniture design. His work has been exhibited in Italy, Germany and Canada as well as in his native Japan. 112

Alfredo Sandoval was born in 1980 in Caracas, Venezuela. He studied industrial design in Caracas, and in 2006 received his MA in industrial design from

Polytechnic School of Design, Milan, Italy. He has returned to Venezuela, where he has founded a collective for young South American designers. 69

Aziz Saniyer was born in Istanbul, Turkey, in 1950. As a child, he was fascinated with carpentry, and later spent his summers working for furniture producers. He founded his own studio, Derin, in 1971 while he was still at college, designing modern furniture and offering an interior design service. In 1981 he set up his own workshop to produce his progressive designs, and in 1999 became partners with his son, Derin, who studied design in Italy. Aziz has now handed over management of the company, which has developed an international reputation, to his son, but he continues to provide technical and artistic assistance. Over the past five years, he has designed products for clients such as Cappellini, Moroso, Zeritalia, Sica and Derin. 30

Richard Sapper, who is based in Milan, Italy, was one of the most influential designers of the twentieth century and has produced award-winning products throughout his career. He is strongly associated with post-War Italian design, and is well known for his Tizio lamp, produced for Artemide. He applies his knowledge of engineering and structure to produce high-tech precision products with a sensual quality. 159

Lela Scherrer was born in 1972 and lives and works in Antwerp, Belgium and Basel, Switzerland. From 1997 to 2000, she was a costume designer for various theatres in Switzerland and Belgium. She studied fashion design at HGK Zurich from 1998 to 2001, and then designed her own clothing collection and was a designer for Dries van Noten in Antwerp. Since 2005, she has been a designer for Wim Neels in Antwerp. Scherrer's work has appeared in several exhibitions, and she has been awarded prizes, including the Swiss Design Prize in 2005. 145

Antonio Sciortino was born in Palermo, Sicily, in 1962 and now lives and works in Milan. He trained as a ballet choreographer, but also experimented with crafting iron objects, and after abandoning ballet, dedicated himself fully to design. 177

Ayala Serfaty was born in 1962 in Tel Aviv, Israel, and studied fine arts at the Bezalel Academy of Art in Jerusalem, graduating in 1984. Afterwards, she undertook graduate work at Middlesex Polytechnic, London, from 1985 to 1987. She is the founder, owner (in partnership with her husband, Albi Serfaty) and sole designer of Aqua Creations, a design and manufacturing studio. Her work has been exhibited in group and solo exhibitions internationally. 153, 163

Michael Sieger is an internationally acclaimed photographer, graphic designer, trade-fair designer and creator. He was born in 1968 and studied industrial design at Essen Polytechnic and at the Münster University for Applied Sciences in Germany. Early in 2003 he established Sieger Design in Sassenberg, Germany, and is a managing partner, together with his brother Christian. The company focusses on industrial design and architecture, design management and graphic design, public relations and marketing. 130, 133, 197

Paul Simmons *See* Timorous Beasties

Michael Sodeau was born in London in 1969 and studied product design at Central Saint Martins College of Art and Design. Following his graduation in 1994, he and two partners set up the design studio Inflate, producing inflatable re-interpretations of traditional products. In 1997, Sodeau left Inflate to set up a design studio of his own with his partner Lisa Giuliani, and they launched their first collection, Twentyonetwentyone, which included woven cane lights, ceramic homewares, rugs and furniture. The success of this collection led to commissions for the London-based jewellers Dinny Hall and the Swedish manufacturer Asplund. In 2005, Sodeau was appointed external art director of Modus and also developed many new projects for the hotel market. His work has been widely exhibited, with shows in London, New York, Paris, Stockholm and Tokyo. 209

Bart Spillemaeckers was born in Belgium in 1965 and received his MA in electro-mechanics in 1989. He received both the Design Plus Award and the Red Dot Award for product design in 2006. He works as a designer for the Belgian company Lithos, which manufactures built-in electrical pushbuttons and outlets. 116

Tom Stables studied at Central Saint Martins College of Art and Design in London and undertakes packaging and product commissions for clients such as Heineken and Hulger. 112

Robert Stadler was born in Vienna, Austria. He studied design at the European Institute of Design in Milan, Italy, and then at ENSCI/Les Ateliers in Paris, where he co-founded the Radi Designers group in 1992. Since 2000 he has also worked independently. In 2002 he received a grant to spend six months in Rio de Janeiro, Brazil. His clients include ACME, Magis and Swarovski and he has exhibited at the Espace Paul Ricard, Fondation Cartier and ToolGalerie in Paris and at Klausengelhorn22 Gallery, Vienna. 73

Mattias Ståhlbom was born in 1971 in Norrköping, Sweden, but now lives in the Södermalm area of Stockholm, where he has been running his own design and architect's office since 2002. He received his degree in interior architecture and furniture design from the University College of Arts, Crafts and Design in Stockholm. He has designed furniture and kitchen utensils for a number of well-known Swedish companies and has also worked in New York. Ståhlbom has received numerous awards and has held exhibitions in a number of venues. He has displayed his products at major trade fairs such as Swedish Style in Tokyo, the Milan Furniture Fair and the Design Fair in Seoul. 164

Philippe Starck was born in Paris in 1949 and trained at the Ecole Camondo in Paris. He has been responsible for interior-design schemes for François Mitterand's apartment and multipurpose buildings such as the offices of Asahi Beer in Tokyo. As a product designer he collaborates with Alessi, Baum, Driade, Flos, Kartell and Vuitton. From 1993 to 1996 he was worldwide artistic director for the Thomson Consumer Electronics Group, and from 1999 to 2000 he finished two central London hotels for the Ian Schrager group. Starck has also designed the interiors of the Royalton and Paramount hotels in New York, the Peninsula Hotel restaurant in Hong Kong, the Teatron in Mexico, the Hotel Delano in Miami, the Mondrian in Los Angeles, and the Asia de Cuba restaurant in New York. 23, 72, 137, 138, 147, 167, 176

Storno is a group of four Berlin-based designers – Henrik Drecker, Katharina Ploog, Davide Siciliano and Sven Ulber – who all met at the University of Art, Berlin, Germany. To date, the group has worked on two exhibitions, for 'DesignMai 2003' and 'DesignMai 2005', product design and graphic design. 100

Studio Demakersvan is a design team comprised of Jeroen Verhoeven, **Joep Verhoeven** and **Judith de Graauw**, all of whom were born in 1977 and met at the Design Academy Eindhoven, The Netherlands. In 2005, they worked on their graduation project as a team, and in the process formed their professional partnership. Their work has been purchased by the Victoria & Albert Museum in London. 47, 84

Studio Job is run by Job Smeets, born 1970 in Hamont, Belgium, and Nynke Tynagel, born 1977 in Bergeyk, the Netherlands. From their native countries they work with a team of assistants on a number of national and international commissions. Studio Job's conspicuous iconic figures are remarkable for their grotesque shapes and highly refined finish. Recently, Royal Tichelaar Makkum, which was established in 1572 and is the oldest company in The Netherlands, introduced Biscuit, a new ceramics collection designed by Studio Job, produced using a new porcelain technique. The collection includes nine different white plates and five centrepieces, based on the artists' characteristic allegorical reliefs, with fairy-tale and fantasy figures. 192–3, 217

Studio JSPR was founded in 2005 by **Jasper van Grootel**, who graduated from the Design Academy Eindhoven under the supervision of Oscar Penya. The Eindhoven-based multidisciplinary design and production firm specializes in interior design, product development, media design and production consultancy. Studio JSPR collaborates with architects, fashion designers, graphic designers and artists, and works in various design fields, ranging from interiors to the production of special tiles and others types of products. It also specializes in manufacturing small- and large-scale customer-designed projects. 211

Sebastian Summa and Hrafnkell Birgisson form a product and exhibition design partnership based in Berlin and Reykjavik. Summa was born in Munich in 1973 and studied product design at the University of Applied Sciences in Potsdam, Germany. Birgisson was born in Reykjavik, Iceland, in 1969 and studied product design in Germany at the Art Academy, Saarbrücken, and the University of the Arts in Berlin. In addition to their partnership, the two also work individually with various design collectives. 97

Satoshi Suzuki was born in Yokohama, Japan, in 1962 and studied industrial design at Ikuei Technical College (now Salesian Polytechnic). He worked for Poppy Co. (now Plex Co.), a design firm serving Bandai Co., where he was involved in designing robots, then for the design firm NID, before joining Sony Corporation. Since then he has been mainly in charge of Personal Audio, with recent design works including the MDR-SA series and MDR-EX 90 headphones. He currently serves as art director, PA Group, in the Mobile Design Studio of Sony's Creative Center. 120

Martin Szekely was born in 1956 and lives and works in Paris. He began his career in 1983, and since then has created a variety of products, from furniture to jewellery, perfume bottles and drinking glasses, and has also worked as an interior designer. Among his clients are Lavigne, Cassina, the Hermès Group, Fusital, Heineken and Dom Pérignon Champagne. His work has been exhibited internationally, and is in the permanent collections of numerous museums, including the Centre Georges Pompidou, Paris, and the San Francisco Museum, USA. 78–9

Ezri Tarazi, born in Jerusalem in 1962, is an Israeli industrial designer and educator. Tarazi studied industrial design at the Bezalel Academy of Art and Design in Jerusalem and established the Tarazi Design Studio, located in Shoham, Israel, in 1996. He was the head of the Department of Industrial Design at the Bezalel Academy from 1996 to 2004 and from 2004 to 2006 chaired the Bezalel master's degree programme in industrial design. He has also written on design issues for a number of publications. His best-known work is the 2005 New Baghdad table produced by Edra of Milan. 84

Raffaele Tedesco *See* Diego Rossi

Thelermont Hupton is a design partnership between David Hupton and Yve Thelermont, who launched the company in 2003 after studying together at the London Metropolitan University since 2000. Their backgrounds are in fine art and furniture-making, and their collections include interior accessories, lighting and tableware. The partnership is involved in one-off projects through to manufactured retail and contract production. 89, 142

Renaud Thiry lives and works in Paris. He graduated from the French National Institute for Advanced Studies in Design, ENSCI/Les Ateliers in 1995 and founded Flandesign in 1995. His clients include Cappellini Progetto Oggetto, Ligne Roset, Cinna, Hermès, Puiforcat, Meiji, Issey Miyake, Habitat, Ladurée and Cartier. Since 2000 he has been a lecturer in industrial design at Lycée Microtechnique Diderot, Paris. His work has been exhibited in France and Italy, including in a solo exhibition at the Biennale Internationale Design Saint-Etienne in 2004, and has appeared in numerous publications. 165, 177

Timorous Beasties was established in 1990 by **Paul Simmons** and Alistair McAuley, who met at Glasgow School of Art, Scotland, in the 1980s. The company name is taken from a line in Robert Burn's poem 'To a Mouse' (pronounced 'moose'). The work of the pair, who have been described by *Blueprint* magazine as 'Textile Mavericks', crosses many disciplines and styles. 215, 218, 219

Tools Design was founded in Copenhagen, Denmark, in 1989 by industrial designers Claus Jensen and Henrik Holbæk. Having received more than 100 awards and distinctions, Tools Design ranks among Denmark's most awarded design groups and has represented Danish design in several international exhibitions. The group's portfolio ranges from electronics and medical equipment to household products. 102, 104

Peter Traag is a furniture designer born in Tegelen in the Netherlands in 1979. He studied 3D design in Arnhem and design products at the Royal College of Art in London. Since graduating from the RCA in 2003 he has been working as a product designer in London, creating his own range of experimental furniture, which has been shown in various exhibitions all over the world. His clients now include Edra, Pallucco and the British Council. 87

David Trubridge, who graduated as a naval architect in the UK, is New Zealand's best-known furniture designer, and has his own design studio and manufacturing workshop at Whakatu, Hawke's Bay. His designs are based on a synthesis of craft knowledge, sculptural abstraction and computer design technology, and incorporate his experience of sailing. He frequently travels, often to wild places, and in 2004 was a visiting artist in Antarctica. Trubridge regularly exhibits in Milan and New York, and his work is manufactured by Italian companies Cappellini, Boffi and Emmemobili. His designs have featured in more than eighty magazines worldwide, and he has also won various New Zealand design and Arts Council Awards. He has spoken at various overseas conferences, including the first Chinese Design Conference in Dongguan. 165

Tsé and Tsé Associates was formed in Paris in 1989 by **Catherine Lévy** and **Sigolene Prébois**, who met while studying at ENSCI/Les Ateliers. Tsé and Tsé creates objects for the home, covering all stages necessary for the development of a product, from design to manufacture, seeking out artisans or manufacturers to implement them and marketing them through a small number of hand-picked distributors in Europe, the USA, Brazil, Asia (Japan, Singapore, Taiwan, etc.), Australia and New Zealand. In parallel with the development of its own products, Tsé and Tsé is currently working on projects for international brands such as Lancôme, Habitat, Issey Miyake, Ricard, Baccarat, Salviati, MK2 cinema and Springcourt. 187

Suzanne Turell *See* Cam Bresinger

Paolo Ulian was born in Massa-Carrara, Italy, in 1961. He studied painting at the Academy of Fine Art in Carrara and industrial design at the ISIA in Florence, graduating in 1990. His work has been shown in numerous exhibitions worldwide and he collaborates with companies such as Driade, Droog Design, Coop, Indarte, View, Sensi&C., Progetti, Fontana Arte, Seccose, Luminara, and Zani & Zani. He has won several awards. 91

Kiki van Eijk was born in 1978 in Tegelen, The Netherlands. She graduated from the Design Academy Eindhoven in 2000, and became well known for her Kiki Carpet. Her work is both two- and three-dimensional, conceptual, functional and often narrative. Among her clients are Studio Edelkoort Paris, Centraal Museum Utrecht, Swarovski, Moooi and Appèl. Van Eijk's work has appeared in numerous publications and is sold worldwide. 81, 201

Niels van Eijk was born in Someren, The Netherlands, in 1970 and studied at the Polytechnic School in Helmond, Tehatex in Nijmegen and the Design Academy Eindhoven. His work is in the collections of several museum collections in The Netherlands and abroad, including the Stedelijk Museum, Amsterdam, Centraal Museum, Utrecht, and Manchester City Art Gallery, UK. 150

Jasper van Grootel *See* Studio JSPR

Naomi Sara van Overbeeke graduated from ABK Maastricht in 2004 and is now employed as a textile designer at Innofa, The Netherlands. 199

Tjeerd Veenhoven studied in The Netherlands at Academie Minerva Groningen (1994–7) and 3D design at Hogeschool voor de Kunsten Arnhem (1998–2002). His work has been exhibited in Italy and Belgium as well as at various locations in his native country and has been featured many times in the international press. 230

Joep Verhoeven *See* Studio Demakersvan

Louisa Vilde was born in 1976 and graduated with a degree in metals and jewellery from Monash University, Caulfield, Australia. Her work has been widely published and exhibited internationally, and has received a number of awards, including the Contemporary Australian Silver and Metalwork Award, Buda, Castlemaine, Australia 2005. 229

Clara von Zweigbergk, who was born in Stockholm in 1970, studied illustration and graphic design in Sweden,

France and the USA. After she graduated, she worked as a graphic designer for several advertising agencies in Sweden and Milan, Italy, for clients such as MTV London, Wella and Nordiska Kompaniet. She is currently a graphic designer for a large design studio in Milan specializing in architecture and furniture, product and graphic design. 129

Voon Wong & Benson Saw, established in London in 2001, develops designs for manufacturers and clients such as Fontana Arte, Porro and Kirin and produces its own range of tableware and accessories. Its furniture, lighting and tableware has been exhibited in London, Cologne, Milan and Singapore. In 2002 Voon Wong & Benson Saw was awarded the OXO lighting prize for its Elma vases, and its Loop lamp was shortlisted for the prestigious Compasso d'Oro 2004. The partnership is also involved in interior design and architecture with current projects in London, Singapore and Shanghai. 86

Roderick Vos was born in The Netherlands in 1965 and studied industrial design at the Design Academy Eindhoven. He worked for Kenji Ekuans GK in Tokyo and Ingo Maurer in Munich before co-founding a studio, Studio Maupertuus, with Claire Vos-Teeuwen. The studio's clients include Espaces et Lignes, Driade, Authentics and Alessi, and its work has been shown at the Milan, Cologne and New York furniture fairs. 161

Jakob Wagner, who was born in 1963, graduated from the Danish Engineering Academy in 1987 and studied industrial design at the Royal Danish Academy of Art (1988–90) and product design at the Art Centre College of Design in Pasadena, California (1991–2). He later opened his own design studio in Copenhagen. His work has been featured in numerous exhibitions worldwide and has received a number of awards. Wagner teaches design courses at the Danish Design School and lectures at international conferences on design topics. 99

Sharon Walsh is a London-based textile designer whose designs are inspired by the concept of creating new from old. Her individual approach to design won her the runner-up prize in the New Designer of the Year Award 2004. 46, 219

Marcel Wanders grew up in Boxtel, the Netherlands, and graduated from the School of the Arts, Arnhem, in 1988. He designs products for the some of the most prestigious contemporary design manufacturers, such as B&B

Italia, Bisazza, Poliform, Moroso, Flos, Boffi, Cappellini and Droog Design, and is the art director and co-owner of Moooi, established in 2000, which has grown into an internationally renowned design label. Additionally, he works on architectural and interior design projects and has recently begun to develop consumer home appliances. Wanders' designs have been exhibited and published worldwide and are in the permanent collections of the Museum of Modern Art in New York, the Victoria & Albert Museum in London, the Stedelijk Museum Amsterdam and many others. He has won numerous international awards, most recently the *Elle Decoration* International Design Award for Designer of the Year (2005 and 2006). 54, 61, 71, 75, 137, 156

Norbert Wangen was born in Prüm, Germany, in 1962 and now lives and works in Lugano, Switzerland, and in Vienna. He studied sculpture at the Kunstakademie Düsseldorf and architecture at Aachen and Munich, and in 1991 graduated in architecture from the Technical University, Munich. In 1995 his patented folding armchair, Attila, was included in the collection of the Neue Sammlung München, and in 1997 in the permanent collection of the Vitra Design Museum in Weil am Rhein, Germany. In 1995–7 he designed the first kitchen with a sliding worktop developed for a residential reconstruction in Munich. Wangen was selected as one of the 100 best architects and interior designers in Germany by the review *Architektur & Wohnen*. In 2003 the Norbert Wangen brand became part of Boffi. 132

Clemens Weisshaar See Kram/Weisshaar

Lee West, born in 1976, is an independent British designer based in Paris. He studied product design at Ravensbourne College of Design and Communication, London, and graduated in 1999. That same year, he won *The Times* New Designers Award for Best New British Designer of the Year for a collection of work exhibited at the 'New Designers 99' show in London. From 2000 to 2004, he collaborated with leading Parisian design agencies, working on various projects ranging from perfume bottles, Champagne service sets, glassware and watches to concept computers. In 2005 he had a solo exhibition at the British embassy in Paris. West's clients include Ligne Roset, Cinna, Arc International, Paco Rabanne and Seiko. 72

Hannes Wettstein was born in Ascona, Switzerland, in 1958. Since 1991 he has been teaching in tandem with

his professional activities, first as lecturer at the Swiss Federal Institute of Technology, Zurich, and then as professor at the Karlsruhe Hochschule für Gestaltung in Germany. In addition to furniture and furnishing accessories, he designs professional audio equipment and watches. His clients include Baleri, Cassina and Bulo. 45, 63, 104

Mary-Ann Williams was born in 1965 in Cape Town, South Africa, and studied fashion design and interior design in Cape Town and Hamburg. In 1984 she started using felts for fashion and hats and later specialized in the development of textile products in the high end of interior design. She creates accessories for the home, including bowls, lights, tableware and kitchen tools, and also greeting cards, bags, carpets, hats and fashion accessories. Her best-known designs are Flokati felt carpets, which she invented, and a machine-washable wool felt. She has received numerous awards, including the iF (International Forum Design) gold award, Hannover. 227

Damian Williamson, who was born in 1974 in London, where he studied at the London College of Furniture and Wimbledon School of Art before obtaining a degree in 3D design at Kingston University. After graduating, he went to work at Sandellsandberg, where he was design assistant to Thomas Sandell until 2003. Since 2001 Williamson's work has featured in numerous exhibitions and publications in both Sweden and Italy, and in 2006 he presented a solo exhibition at the Stockholm Furniture Fair. In 2004 he founded Damian Williamson Office for Design in Stockholm, Sweden, where his client list has included David Design, Gärsnäs, BAS Brand Identity, FORUM and Stockholm International Fair. Since 2005 he has served as guest critic at the LTH Design University Sweden. 32

Wilmotte & Associés, based in Paris, was founded by architect, town planner and designer Jean-Michel Wilmotte in 1975. Today its team of more than 100 people from various nationalities works on more than 100 projects in France and abroad. Activities range from industrial design to architectural projects. 155

Sebastian Wrong studied sculpture before founding his own manufacturing company. He has been in the manufacturing sector for the past ten years, and has collaborated with a diverse selection of designers and firms. In 2002 his Spun lamp caused a sensation in the

design world, and it subsequently won the Red Dot design award and is now produced by Italian lighting manufacturer Flo. In 2003, Wrong and Mark Holmes formed The Lane, which has produced a collection of design products and exhibition designs. Wrong is a founding member and operations director of the design firm Established & Sons. 26

Naofumi Yoneda graduated from the Rhode Island School of Design, Providence, in 1993 and joined Sony the same year. His product designs range from personal audio equipment to video cameras and computers. Among his most important designs are the MDR-V7000 DJ headphones, the VAIO QR, 52 Sports Art Direction and the XDR-M1 portable DAB radio. 113

Tokujin Yoshioka was born in Japan in 1967 and studied at the Kuwasawa Design School in Tokyo with Shiro Kuramata and Issey Miyake, graduating in 1986. His enduring collaboration with Issey Miyake, which began in the mid-1980s, has resulted in various projects, including the shops Issey Miyake and A-Poc. Yoshioka began working as a freelance designer in 1992 and established his own studio in Tokyo in 2000. He has received several prestigious awards for his experimental designs, including the JDC Design Award of Excellence in 1999 and the Mainichi Design Award in 2001. His Honey-pop paper armchair (2000) is in the permanent collections of the Museum of Modern Art in New York, the Centre Georges Pompidou in Paris, and the Vitra Design Museum in Berlin. 63, 79, 87, 89, 118, 148

Michael Young has been among the most successful and influential designers of his generation from the outset of his career. Born in Sunderland, England, in 1966, he studied furniture and product design at Kingston University in London. After graduating, he worked for four years for Space, the design studio run by Tom Dixon. In 1994 he opened his own studio in London and a second office in Reykjavik, Iceland. Young has developed products and furniture for such manufacturers as Artemide, Cappellini, Danese, Rosenthal and Swedese as well as interior and architectural projects. He divides his time between Reykjavik and his studio in Brussels. 126

Photographic credits